Environmental Management Systems Handbook for Refineries

Environmental Management Systems Handbook for Refineries

Pollution Prevention Through ISO 14001

Nicholas P. Cheremisinoff, Ph.D.

Houston, Texas

Copyright © 2006 by Gulf Publishing Company, Houston, Texas.
All rights reserved.
No part of this publication may be reproduced or transmitted in any form without the prior written permission of the publisher.

Gulf Publishing Company
2 Greenway Plaza, Suite 1020
Houston, TX 77046

10 9 8 7 6 5 4 3 2 1

Printed in the United States of America.

Printed on acid-free paper.

Text design and composition by Ruth Maassen.

Library of Congress Cataloging-in-Publication Data

Cheremisinoff, Nicholas P.
  Environmental management systems handbook for refineries : pollution prevention through ISO 14001 / Nicholas P. Cheremisinoff.
    p. cm.
  Includes index.
  ISBN 0-9765113-8-X (alk. paper)
  1. Petroleum refining—Environmental aspects. 2. Petroleum refineries—Design and construction. 3. Environmental protection. 4. Environmental chemistry. 5. Environmental management. I. Title.
  TD899.P4C475 2006
  665.5'30684—dc22
                        2005031495

Table of Contents

About the Author ix
Preface xi

1 Is Compliance Enough? 1
  Introduction 1
  Environmental Cost Accounting 3
  Liabilities for Simply Meeting Minimum Compliance
    Requirements 26
  Understanding the Elements of an EMS 48

2 Getting an EMS Started 55
  Introduction 55
  Practical Aspects of an EMS 56
  Examples of EMSs 65
  Defining Pollution Prevention and Related Terminology 74
  Pollution Prevention within the Context of an EMS 80
  The Initial Environmental Review 85

3 Integrating Cleaner Production with an EMS 111
Introduction 111
The Link between Pollution Prevention and an EMS 113
Integrating EMS and CP/P2 115
The Role of an Environmental Management Information System 121
Recommended References 128

4 Improving Competitiveness through an EMS 129
Introduction 129
Benefits That Improve Competitiveness 129
A Case Study on Greening the Supply Chain 135
Greening the Supply Chain 140

5 Conducting Cleaner Production Audits 147
Introduction 147
In-Plant Assessment and Pollution Prevention 148
The Environmental Management Hierarchy 151
Reasons to Prevent Pollution 153
Improving Product Quality through Total Quality Environmental Management 154
Steps to Conducting CP/P2 Audits 158
Toolkit and Worksheets 190

6 IER Case Study 207
Introduction 207
The Methodology Applied 209
Environmental Management System Structure 213
The Initial Environmental Review 213
Final Remarks 258

7 Cleaner Production Case Studies 261
Introduction 261
CP/P2 Case Study at the Jordan Petroleum Refinery Company, Zarqa, Jordan 262

CP/P2 Case Study at the Chevron Products Company, Richmond Refinery, United States 293
CP/P2 Case Study at the Beijing Yanshan Chemical Corporation, China 296
CP/P2 Case Study at the Indian Oil Corporation Limited, Gujarat Refinery, Vaodara, India 299
CP/P2 Case Study at California Refineries 303

Index 307

About the Author

Nicholas P. Cheremisinoff has 30 years of international consulting and project management experience across a broad spectrum of industry sectors dealing with environmental management practices and technologies. His assignments have focused on corporate responsible care through the application of pollution prevention and cleaner production. He has served as both a practitioner and an educator to industry stakeholders on responsible care practices that strive to improve environmental performance through the application of cost-effective clean technologies. He designed and implemented projects for the United States Agency for International Development, the World Bank Organization, the European Union, the U.S. Trade and Development Agency, the U.S. Department of Energy, the U.S. Department of Defense, and a number of Fortune 100 and multinational companies. He has applied his experience and knowledge in major environmental toxic tort litigations as an expert witness in the fields of environmental forensics, historical reconstruction of environmental management practices, and levels of responsible care exercised by industry stakeholders. He is the author, coauthor, or editor of more than 150 books and numerous state-of-the-art review articles and R&D papers. He received his B.S., M.S., and Ph.D. degrees in chemical engineering from Clarkson College of Technology.

Preface

This volume is intended as a primer for implementing an environmental management system. The focus is on refinery operations; however, the principles and general discussions are applicable to all manufacturing sectors.

All business operations are faced with global challenges involving dynamic interactions between changing regulations, market competitiveness, and spiraling costs for raw materials and labor. It is impossible to separate environmental management issues and infrastructure investments from the primary strategies of any business. In fact, any company, facility, or refinery that focuses on pollution control as opposed to environmental management will not find itself in business over the next decade. There simply are too many costs and financial exposures to businesses that do not focus on continual improvement of their environmental performance. A commitment to continual improvement of environmental performance is increasingly becoming the standard of care for industry.

While the printed literature and the World Wide Web abounds with information on environmental management systems (EMS) and pollution prevention, few publications treat the two subjects in a single volume. This is odd, because ISO 14001, the universal EMS, is based on commitment to pollution prevention or cleaner production practices and technologies. A few years ago the book *Green Profits* (my earlier publication) represented one of the first publications that attempted to link these two subjects more closely on a practical level. This volume relies much on the principles presented in that earlier volume but relates it to refinery practices in later parts of this volume.

The volume is not meant to be all-inclusive, nor does it attempt to summarize the vast literature on the application of EMSs. Instead, the intent is to supplement the literature with a focus on one industry sector. The book's usefulness is enhanced with a CD containing a toolkit of audit forms that can be used for applying a pollution prevention audit. As explained in the book, pollution prevention (P2) practices are the central focus of any EMS and the basis by which companies achieve continual improvement to environmental performance.

The book contains seven chapters. Chapter 1 provides case studies that emphasize the absolute necessity for facilities and businesses to maintain an EMS with P2 at its core. A dedicated EMS is simply a matter of responsible care, which in turn has an impact on a business's sustainability. Chapter 1 also provides an explanation of what an EMS is, how it differs from pollution control, and orients the reader by further drawing attention through case studies that illustrate that meeting minimum compliance requirements is inadequate by today's standards.

Chapter 2 covers the elements of an EMS in greater detail, with a focus on the initial environmental review, or IER. This is a technique and tool to assist a refinery in identifying and prioritizing the most important environmental issues that should be addressed first.

Chapter 3 provides a management model that explains more clearly how pollution prevention or cleaner production practices and technologies are intended to work and enhance an EMS. This chapter also contains a discussion on environmental management information systems (EMIS). An EMIS is an essential tool by which the EMS can be effectively disseminated throughout the refinery and business as a whole.

Chapter 4 is a quick summary chapter that provides a link between the need for an EMS and sustainability and the financial wellness or competitiveness of a business unit.

Chapter 5 provides a stepwise approach to conducting pollution prevention/cleaner production audits. The chapter is compiled in a concise, prescriptive manner, so that pollution prevention teams can cut directly to the practical aspects of the audit and begin implementation. The CD contains a toolkit that will facilitate the audit.

Chapter 6 is a detailed case study of an IER conducted at a small refinery in the Middle East. The methodology applied provides an example of how early savings and benefits can be achieved with an integrated EMS/P2 program.

Finally, Chapter 7 provides select case studies of cleaner production practices at refineries.

I wish to acknowledge and thank the Jordan Petroleum Refinery Co. for its assistance and willingness to work with me and in making a major commitment to implementing a formal EMS. A special thanks is extended to Gulf Publishing for the fine production of this volume.

Nicholas P. Cheremisinoff, Ph.D.

1
Is Compliance Enough?

Introduction

This first chapter addresses this question: Is compliance enough? If your business and manufacturing operations have a good or reasonably good track record in terms of complying with current environmental regulations and standards, then the company on the whole must be doing a good job in terms of its environmental performance. Or is it?

Less than a decade ago, several books appeared on the scene that expounded the need for going beyond compliance. Most of these focused on a core principle that is explored later in this book, namely, that exceeding minimum guidelines can enhance a business's financial bottom line through indirect benefits, such as achieving a competitive advantage through a positive corporate image as a "green" business. This and other benefits discussed later on are all true; however, there is even a greater reason—sustainability.

Sustainability has two meanings within the context of an environmental management system (EMS). The ability to preserve and stretch resources is achieved through pollution prevention and dedicated source reduction programs; waste minimization and recycling enable companies to stretch finite resources, thereby ensuring resource availability for future years of operation. This concept of sustainability is discussed in Chapter 4.

The second interpretation of sustainability is in terms of future risks to a business. In this regard, we borrow a term from the legal profession: liability. All companies, regardless of the industry sector within which they operate or how good their compliance records are, face future liabilities. What is a well-managed waste problem in today's context may not be in the future. This is often thought of by corporations as an intangible cost for environmental compliance and is generally discounted heavily by captains of industry and business leaders. This is a mistake, for many corporations that acted within the existing legal environmental framework for many years subsequently faced huge liabilities that threatened the very existence of their businesses. Many of these examples are not widely publicized; hence, industry-sector specific examples are often difficult to provide in a general publication. However, the book includes a few examples where I was involved as an environmental forensic expert in the reconstruction of manufacturing events and their relationship to responsible care issues on the part of companies. These will help show why the intangible should be considered substantive in the long-range planning for the sustainability of any business operation.

While the focus of this volume is on the petroleum industry, a refinery and its downstream and upstream operations cross other sectors, and in terms of generalized concepts, refineries do not rely on unique business practices that set them apart from other industries. In short, from a general standpoint, the principles discussed in this first chapter, and in fact through Chapter 5, are equally applicable to all industry sectors.

As noted in the Preface, it is not the intent of this book to review in detail all the concepts, practices, and procedures associated with an EMS. The literature is rich, diverse, and extensive: almost a limitless number of resources are available both in print and on the World Wide Web (WWW), which the reader can access for the operational aspects and implementation practices of an EMS. Some of these excellent resources are referenced in various chapters, but none stands alone as an all-inclusive, authoritative source. This volume attempts to supplement an already rich and diverse technical literature with concepts and methodologies that will help a company shape its strategy in meeting several essential elements of any business—sustainability, competitiveness, and profitability—through improved environmental performance.

Environmental Cost Accounting
Introducing the Concepts

Financial calculation tools applied to the analysis of investment decisions and life-cycle costing (LCC, also life-cycle costs) are closely related techniques for evaluating investments involving expenditures for equipment, installation, services, and various infrastructure investments needed to maintain and grow a business. These investments have present and future benefits or will have an impact on future costs. Readers familiar with business financial management will recognize the process of evaluating and selecting from among investment alternatives having the same objectives as *capital budgeting*. Among engineers this process is referred to as *engineering economic analysis*. Whatever the term, the process of economic evaluation and comparison of investment alternatives encompasses a collection of analytical techniques that are tools for investment decision making. For environmental management, these tools enable a company to assess the financial benefits of alternative strategies or compare such strategies to the status quo. Ideally, investments in terms of costs, cost savings, and potential for reducing risk of environmental liability are the foundation on which smart companies base their long-term strategies, just as they do for any business. Unfortunately, there still exists a mentality among enterprises to manage the environmental issues separate from the overall business planning and strategic investments, which are the mainstream of the business. This is mistake, because in today's world it is not possible to separate the mainstream business from that business's environmental obligations.

The essence of any investment is to sacrifice now in favor of future benefits. A typical investment decision asks this fundamental question: Do the future net benefits from the investment outweigh the initial costs? Benefits may be in the form of additional future income or revenues, or they may be intangible and possibly nonpecuniary in nature.

LCC addresses the question of how best to accomplish a particular task. In other words, it assists in defining the least-cost method, taking into consideration both the initial outlay and future operating costs. LCC also involves risk assessment, in the sense that some of the strategies or initiatives under consideration require larger initial outlays and achieve lower future costs than others. In developing criteria for an investment portfolio, a business should focus on the direct costs for investments and the potential for reducing long-term liabilities. In terms of the environmental issues facing companies, this requires that attention be given to the several cost tiers.

Decisions on investing in pollution management strategies that rely heavily on end-of-pipe treatment technologies are the result of ignoring or not recognizing the true or total costs associated with pollution control and waste management. To account for all of the costs we need to recognize that they fall into four distinct categories or tiers (examples of each category follow):[1]

1. Usual or normal costs.
2. Hidden or direct costs.
3. Future liability costs.
4. Less tangible costs.

Tier 1 Costs. Usual or Normal Costs

These costs include such costs to the operation as

- Pollution fees (i.e., fees paid for discharging pollutants within legally allowable emissions standards).

---

[1]D. McHugh, *The Economics of Waste Minimization* (New York: McGraw-Hill Book Publishers, 1990).

- Direct labor costs.
- Raw materials (e.g., feedstock chemicals, water).
- Energy.
- Capital equipment items.
- Site preparation for pollution control equipment.
- Equipment tie-ins and process modifications to accommodate pollution controls.
- Employee training.
- Permits to construct.

Tier 2 Costs. Hidden or Indirect Costs
These include such costs to the operation as

- Expenses associated with monitoring.
- Permit fees (e.g., permits to operate, permit renewals, other).
- Environmental transformation costs. (For example, costs are associated with transforming a water pollution problem into a solid waste disposal issue; an example is a dewatering operation such as a filter press to treat suspended solids in a wastewater stream. This strategy generates a sludge that requires disposal. Another example is a wet scrubber used to capture particulates to address an air pollution problem. In this example, although we eliminate or reduce an air emission problem, we generate a water pollution problem with the use of this technology. The water pollution problem may also have to be addressed by separate treatment technologies, resulting in a sludge disposal problem.)
- Environmental impact statements (EIS; depending on the nature of the project, EISs can cost many thousands of dollars and require long lead times).
- Health and safety assessments.
- Service agreements.
- Legal costs.

- Control instruments.
- Maintenance and replacement costs.
- Reporting and recordkeeping.

Tier 3 Costs. Future Liability Costs
These include such costs to the operation as

- Remedial action costs (site cleanup plus cleanup costs associated with third-party damages).
- Personal injury to workers.
- Health risks and injury to the public.
- More stringent compliance (environmental compliance is oftentimes a moving target, hence, a control technology used today may become obsolete a few years down the road).
- Inflation (e.g., this could have an impact on higher feedstock costs and energy as well as waste disposal services).

Tier 4 Costs. Less Tangible Costs
These include such costs to the operation as

- Consumer response and loss of confidence due to perceived poor environmental management (it is well recognized in technologically advanced societies that consumers and investors favor those companies that are environmentally conscious of their actions and have proactive environmental management programs).
- Employee relations (poor management of environmental issues places workers at health risks).
- Establishing and extending lines of financial credit (lending institutions will not extend loans to or provide favorable terms for credit lines if a company runs a high-risk operation from an environmental management standpoint).
- Adverse impacts on sales of property or mergers and acquisitions (companies that are looking to partner or merge with

others are not willing to take on the other firm's environmental liabilities; poor environmental management can bring enormous liabilities to investors and partner organizations).

- Higher insurance premiums (companies that have a poor environmental track record represent a risk and, therefore, the potential loss of assets from fire or explosion or health risks to workers and the public; insurers will not take these risks, or they will impose high premiums with very limited coverage).
- Becoming the focus of frequent environmental inspections and corrective actions (if environmental regulators recognize a recurring theme of noncompliance or accidents, then the facility will be singled out for numerous violations, fines, inspections, and corrective actions, all of which can result in significant costs to an operation, including interruptions in production schedules and legal costs).
- Less leverage in negotiating environmental fines and penalties. (The environmental laws in technologically advanced nations are complex and operations are oftentimes fragmented, leading to innocent violations of environmental laws. If a company has had a history of poor environmental management, then it has little basis to negotiate for leniency on innocent violations. The problem is perceived as systematic.)

Tier 1 and 2 cost components are usually identified in selecting pollution management strategies. If we look at these tier cost components alone, then it is easy to justify investments into control-based technologies. In contrast, tier 3 and 4 costs are often not given a sufficient weight (i.e., they are considered less important than more direct and tangible costs, usually because components are more difficult to assign values to), or they are simply not recognized. This is a mistake, because by attaching the proper importance and most probable costs to these components, other strategies become apparent and more attractive.

Proper evaluation of tier 3 and 4 cost components in assessing total costs and applying the results as a basis to selecting proper pollution management strategies requires the application of risk assessment principles. Since the ultimate goal is to arrive at cost-effective strategies that meet (or go beyond) a compliance target, risk assessment should be considered a part of life-cycle costing analysis (LCCA).

Understanding and dissecting the cost components within each tier are critical to making decisions on investing into the most economical pollution management strategies. When companies focus their attention on the obvious costs (tiers 1 and 2), critical factors that could alter the decision on investments into pollution prevention may be missed. As an example, consider an electroplating operation that relies on the use of chromium solutions. This type of operation produces highly toxic sludge and wastewaters that contain hexavalent chromium, a confirmed human carcinogen. Stringent environmental regulations demand the applications of best available technologies (BAT) to control wastewater emissions and dispose of the toxic sludge in a safe manner. But BATs generally tend to be expensive, especially for regulated hazardous wastes.

By focusing only on tier 1 and 2 cost components, it is quite possible, depending on the size of the operation, that the least-cost strategy is to continue with BAT (i.e., end of pipe). When the analysis focuses on such issues as worker health risks, cumulative sewer discharge fees, on-site wastewater treatment costs, and the cumulative costs associated with sludge stabilization, on-site staging of wastes, recordkeeping and waste reporting requirements, and transport and disposal, then other alternatives that displace or eliminate the generation of hexavalent chromium wastes as a part of the process become much more attractive. When we consider also the potential future liabilities when the facility is sold, such as the costs that may be faced in remediating the property, then higher investments into strategies that eliminate chrome as a part of the operation are justifiable. This is the basis under which life-cycle costing techniques are applied to making logical investment decisions.

Table 1–1 provides a summary of the cost/benefit tiers.

Table 1-1 Pollution Cost Tiers

Tier	Cost Category	Typical Cost Components
1	Usual or normal	Direct labor Raw materials Energy Capital equipment Site preparation Tie-ins Employee training Permits
2	Hidden or direct	Monitoring Permitting fees Environmental transformation Environmental impact assessments Health and safety assessments Service agreements Legal Control instrumentation Maintenance and replacement Reporting and recordkeeping
3	Future liabilities	Remedial actions Personal injury Health risks and injury to the public More stringent compliance requirements Inflation
4	Less tangible	Consumer response and loss of investor confidence Employee relations Establishing and extending lines of credit Property values Insurance premiums Frequent environmental inspections and penalties Leverage in negotiating with environmental regulators

Tier 1 and 2 costs can be handled in a normal LCC calculation to identify the least-cost initiatives when compared to a base case or among themselves. With the exception of inflation, tier 3 and 4 cost/benefits must be addressed in other ways. These are best accounted for in the overall investment analysis by assigning levels of risks to supplement investment decisions. The levels of risk of certain liabilities or future events can be determined based on probabilities, which in turn can be qualified by the application of confidence limits.

An example of this is a company responsible for operations requiring the management of a large number of USTs (underground storage tanks) that handle hazardous and regulated materials. A review of company historical records and industry trends may reveal the enormous costs associated with remediating contaminated groundwater, plus the legal liabilities associated with off-site property damage and litigations stemming from leaking USTs. If a significant number of the tanks are still single-walled vessels or are not equipped with modern leak detection and cathodic protection technologies, overfill containment capabilities, or have old dispensing and transfer lines, then the company should assign a high probability for encountering the various costs and liabilities under the tier 3 and 4 cost categories. The company might quantify this probability with a numerical value (e.g., 75% chance for incurring tier 3 or 4 costs and liabilities for certain operations or facilities in the company), and further qualify the likelihood by assigning a level of confidence (e.g., 65% confident that, over five years or more, leaking USTs could result in the liabilities and costs). The basis for establishing confidence limits can be developed from subjective criteria. Examples of subjective criteria in this case include

- Industry data reflect that 90% of buried single-walled steel vessels more than 15 years old leak.
- The company has had several leaky USTs in the last five years with remediation costs exceeding several million dollars.
- Inventory balances from sales and purchases of the chemicals stored cannot be resolved on a consistent basis.

- Facilities are close to densely populated areas and pose high risks.

By incorporating subjective risk assessment principles to supplement LCC calculations, a company can rank the financial risks to its business and devise strategies and plan investments aimed at reducing the environmental exposures created by its operations. In the UST example, the various strategies might include

- Investing in cathodic protection technologies for some percentage of the USTs that are older than 10 years.
- Implementing automated inventory gauging and interfacing data with a centralized computer as an early warning system for leaks.
- Modernizing all USTs over the next several years.

A business can carry the analysis further by examining integrated systems approaches, which would provide options for implementing several initiatives at once or staging them over a period of time. This enables management to make long-term business growth decisions and is based on life-cost planning (LCP) techniques.

LCP concerns the assessment and comparison of options and alternatives during the design and acquisition phases of a project. It considers all cost components within an asset's options over the asset's life. It does not directly consider the benefits or revenue streams generally assumed to be equal among the options being compared (benefits and revenues are considered in the evaluation of options). LCP and how it can be effectively applied to pollution prevention (P2) investments is a subject that requires a separate volume. Suffice it here to say that the basic concepts used for long-term business financial planning purposes are applied through LCP.

Life-cycle costing tools fall into two general categories: standard LCC calculations and supplemental measures or indicators of economic performance. In performing the analysis it is important to bear in mind that consideration must be given to the time value of money (TVM). This refers to the fact that money in hand right now could be invested elsewhere (in something that has a positive rate of interest). That money with its accumulated interest would amount to more in the future than it does right now. As such, any money to be received or spent in the future is equivalent to a smaller sum of money to be received or spent right now.

The type of initiative sets the pace for the analysis. Costs associated with any investment initiative include

- Initial investment costs.
- Operation and maintenance costs.
- Energy and water costs.
- Residual values.
- Financing costs.

Life-cycle cost analysis is an economic method of project evaluation in which all costs arising from owning, operating, maintaining, and disposing of an investment are considered germane to the decision. The tools are particularly well-suited to the evaluation of design alternatives that satisfy a defined performance level but may have differing investment, operating, maintenance, or repair costs, even possibly different life spans. LCCA tools can be applied to any capital investment decision. They are especially relevant in situations where high initial costs are traded for reduced future costs; as such, LCCA is most relevant to evaluating moderate- to high-cost investments that eliminate or reduce future liabilities associated with current environmental management practices.

LCCA can be used to compare the costs of existing equipment over a given time period with the costs over the same time period of several alternative P2 measures. All the costs for performing a feasibility study—including technology evaluations involving plant trials, setting up, and financing the P2 investment—can be included in the LCCA, as well as accounting for any offset costs or credits received from the no-cost or low-cost P2 initiatives already implemented.

Individual P2 initiatives can be bundled together to optimize materials savings and improve efficiency, costs, and the environmental performance benefits of a project. Various P2 investments that save significant amounts of energy, improve environmental performance, reduce greenhouse gas emissions, improve productivity, save on materials and water, or improve product quality may be bundled with other P2 investments, so long as the overall project is cost-effective in life-cycle terms. All items in the bundle must be complementary: they must be an integral part of the project, and no single P2 initiative should be significantly cost-ineffective.

For P2 initiatives to be practical, they must be more financially attractive than the current situation. In some situations pollution control (end-of-pipe) technologies are the simplest, most cost-effective strategies, especially if a company calculates that the probability of long-term liabilities from disposal practices are low or there are low probabilities for tier 3 and 4 costs.

How an LCC Calculation Works

LCC adds all the costs of alternatives over their lives, enabling evaluation on a common basis for the period of concern. This is usually done using *discounted costs*. This approach enables decisions on acquisitions or disposal to be made in light of full cost implications.

In developing estimates for LCCA, a breakdown of the assets into individual cost elements over time is needed. The level to which the cost elements are broken down depends on the purpose and scope of the LCC analysis, but generally requires identification of the following three elements:

1. Components of the activity that are major cost generators.
2. The point in the life cycle when the work or activity is to be performed.
3. Relevant resource cost categories (e.g., labor, materials, energy, overhead, transportation, etc.).

Costs associated with LCC elements can be further allocated between recurring and nonrecurring costs, and they may also be estimated in terms of fixed and variable costs.

In the following discussion, the formulae and calculation methods used to make an LCC calculation and the need for supplementary economic measures are presented. Complex systems, especially those based on an integrated systems approach, are best handled by LCCA models.

Discounting and Inflation

It is important that the same discount rate and inflation treatment be used in the LCC analysis of alternatives and multiple projects. Project-related costs occurring at different points in time must be discounted to their *present value* at the start date before they can be combined into the LCC estimate for that project. The discount rate used to discount future cash flows to present value is based on the TVM. The discount rate is based on the minimum acceptable rate of return (MARR) for investments of equivalent risk and duration. The MARR is somewhat subjective and depends on how conservative or risk averse individual companies and managers are.

Interest, Discounting, and Present Value

When faced with choosing among P2 investments, top management should be sensitive to the timing of cash flows or savings generated by each investment. In general, it is considered preferable to receive or save a dollar sooner rather than later. There are two reasons for this. First, dollars generally loose purchasing power over time due to inflation. Second, cash amounts received earlier can be reinvested earlier, earning additional returns (another reason why early P2 savings from no-cost or low-cost P2 initiatives are so important).

When we invest a cash amount at a given interest rate, the future value of that cash amount at any point in time can be computed using compound interest. Consider that an initial sum of P_0 dollars is invested for t years at a rate of interest i, compounded annually. In one year, the yield would be iP_0, which when added to the principal, P_0, gives

$$P_1 = P_0 + iP_0 = P_0(1 + i). \tag{1.1}$$

The future compound amount after t years would be

$$P_t = P_0(1 + i)^t. \tag{1.2}$$

Discount rates are essentially a special type of interest rate that makes an investor indifferent between cash amounts received at different points in time. An investor would just as soon have one amount received earlier as the other amount received later.

Calculations involving discounting are identical to those of compound interest. The discount rate, d, is used like the interest rate, i, to determine the present value (PV) of a cash amount received or paid at a future point in time. The PV of a future amount received at the end of year t, F_t, is calculated as follows:

$$PV = F_t/(1 + d)^t. \tag{1.3}$$

Note that costs that occur at different points in time over the life of the investment cannot be used directly in the LCC calculations, because the dollars spent at different times have different values to the investor. These costs must first be discounted to their present-value equivalent amounts. Only then can the costs be added into the analysis to yield a meaningful LCC that can be compared with the LCC of other P2 investments or the current situation.

The Importance of Inflation

Inflation reduces the purchasing power of the dollar over time (conversely, deflation increases it). When future amounts are stated in actual prices as of the year in which they are expected to occur, they are said to be in *current dollars*.

Current dollars are dollars of any one year's purchasing power, inclusive of inflation. In other words, they reflect changes in purchasing power of the dollar from year to year. In contrast, *constant dollars* are dollars of uniform purchasing power, exclusive of inflation. Constant dollars reflect what the same good or service would cost at different times if there were no change in the general price level (i.e., no general inflation or deflation). In other words, there is no change in the purchasing power of the dollar.

The discounting of future cash flows to present value is not the same as adjusting future costs for general inflation. Even when costs are expressed in terms of constant dollars, they must be discounted to reflect the TVM, which is usually greater than the rate of inflation. A discount rate used with constant-dollar amounts is different from the discount rate used with current-dollar amounts. A real discount rate (net of general inflation) should be used with constant-dollar amounts. A nominal discount rate (inclusive of general inflation) should be used with current-dollar amounts.

Cost Categories

There are several ways of classifying the cost components in LCCA. The most important ones to distinguish among are

- Investment-related and operational costs.
- Initial and future costs.
- Single costs and annually recurring costs.

LCC for P2 includes both investment costs and operational costs. The distinction between the two is most useful when calculating economic measures such as the savings-to-investment ratio (SIR) and the adjusted internal rate of return (AIRR). These economic measures evaluate savings in operation-related costs, which generally are most critical for tier 1 and 2 components. The SIR and the AIRR are measures that evaluate savings in operation-related costs with respect to capital investment costs. This distinction does not affect the LCC calculation itself, nor will it cause an investment alternative to change from being cost-effective to non-cost-effective or vice versa. However, it may change its ranking relative to other independent projects when allocating a limited capital investment budget.

When a company considers these elements, it should consider all acquisition costs, those associated with planning, design, purchase, and construction—investment-related costs. It should also take residual values (based on the value in place, resale value, salvage value, or scrap value net of any selling, conversion, or disposal costs) and capital replacement costs into account in the investment calculus. Capital replacement costs are incurred when replacing major systems or components (e.g., a major upgrade on aeration tanks in a wastewater treatment plant), and they must be paid for from capital funds. Operating, maintenance, and repair (OM&R) costs should include energy and process water costs. These are operational costs. Replacements that are related to maintenance or repair are also considered OM&R costs, not capital replacement costs. OM&R costs are best paid for from an annual operating budget and not from capital funds.

The distinction between initial investment costs and future costs is most appropriate when calculating simple or discounted payback indicators. Costs incurred in planning, design, construction, or the acquisition phase of a P2 investment can be classified as initial investment costs. These costs occur before a system or piece of equipment is placed into service. Costs that arise from the operation, maintenance, repair, replacement, and use of the equipment or system during the life of the operation are future costs. Residual values at the end of the system life, or at the end of the study period for the evaluation, are also future costs.

This brings us to single costs versus annually recurring costs. The distinction between these two determines the type of present-value factor most appropriate for discounting future cash flows to present value. Single costs are one-time costs. They occur at one or more times during the evaluation period (life of the investment) at nonannual intervals. Examples of single costs are initial investment costs, and replacement costs scheduled at intervals longer than one year. Repair costs may also be considered single costs. The proper discount formulas to use in LCC are given in Table 1–2. In this table are included subformulas that can be used to compute a corresponding discount factor. The computed discount factor is a number by which an amount is multiplied to obtain a present value. For discounting one-time amounts to present value, the single present value (SPV) factor is used.[2]

---

[2] S. K. Fuller and S. R. Petersen, *Life-Cycle Costing Manual for the Federal Energy Management Program*, NIST Handbook 135 (Washington, DC: U.S. Department of Commerce, 1995).

Table 1–2 Present-Value Formulas and Subformulas Using Discount Factors for LCCA

Formula Application	Formula	Subformula with Discount Factor
PV formula for one-time amounts: the SPV factor is used to calculate the PV of a future cash amount occurring at the end of year t, F_t, given a discount rate, d.	$PV = F_t \times 1/(1 + d)^t$	$PV = F_t \times SPV_{(t,d)}$ *Note:* Use standard discount tables for values of the SPV factor. As an example, the SPV factor for $d = 3\%$ and $t = 15$ years is 0.642.
PV formula for annually recurring uniform costs: the UPV factor is used to calculate the PV of a series of equal cash amounts, A_0, that recur annually over a period of n years, for a given d.	$PV = A_0 \times \Sigma 1/(1+d)^t$ $= A_0 \times \{[(1+d)^n - 1]/d(1+d)^n\}$	$PV = A_0 \times UPV_{(n,d)}$ *Note:* Use standard discount tables for values of the UPV factor. As an example, the UPV factor for $d = 3\%$ and $n = 15$ years is 11.94.
PV formula for annually recurring nonuniform costs: the UPV* factor is used to calculate the PV recurring annual amounts that change from year to year at a constant escalation rate, e (i.e., $A_{t+1} = A_t \times (1+e)$), over n years, given d. Note that the escalation rate can be positive or negative.	$PV = A_0 \times \Sigma[(1+e)/(1+d)]^t = A_0[(1+e)/(d-e)]\{1 - [(1+e)/(1+d)]^n\}$	$PV = A_0 \times UPV^*_{(n,d,e)}$ *Note:* Use standard discount tables for values of the UPV* factor. As an example, the UPV* factor for $e = 2\%$, and $n = 15$ years is 13.89.

Annual recurring costs are defined as amounts that occur regularly every year during the life of the investment in approximately the same amount or in an amount expected to change at a known rate. The types of costs that fall into this category are energy related, water, and routine annual maintenance. The appropriate present value factor to use with the subformulas given in Table 1–2 for annually recurring costs is the uniform present value (UPV) factor or the uniform present value factor modified for price escalation (UPV*). If the annual amounts are expected to change at a known rate, the UPV* factor should be used for the present value factor.

Timing Considerations and Cash-Flow Diagrams

LCCA requires that project-related costs be accounted for in terms of both their amounts and when they occur. This can complicate an LCC analysis, so the more-common practice is to rely on simplifying or approximating models rather than attempt to estimate the exact timing of all costs. For example, all costs occurring at different times during a year may be treated as occurring at the same time each year.

Another useful technique for dealing with time in LCC is to create a cash-flow diagram. A cash-flow diagram for a P2 investment is a graphic representation of all relevant costs and their timing. A horizontal time line represents the study period and marks each year and key dates (e.g., start date, dates of regularly schedule maintenance, dates of expected replacement costs, and end of useful life). Years can be indicated in terms of calendar years or elapsed years from investment inception. There is no standard convention for cash-flow diagrams, however the usual practice is to show positive costs above the time line and negative costs (e.g., residual values) below the time line.

How to Treat Revenues

LCCA is most appropriately used to evaluate the relative costs of design alternatives or P2 investment options that can satisfy certain expectations (specific environmental performance targets); it is not generally appropriate for evaluating the cost-effectiveness of alternative revenue-generating projects. For example, if LCC were carried out on alternate building designs constructed to produce rental income, LCCA would not be appropriate. The same principle applies if a company considers off-site recycling markets in analyzing P2 alternatives. The potential attractiveness of these kinds of revenue-generating strategies are most appropriately evaluated using benefit-cost analysis (BCA) and return-on-investment (ROI) indicators, which can supplement the LCC analysis. This is not necessarily a hard-and-fast rule. If there are small differences in revenue between one design alternative and another, then it may be appropriate to include them in LCCA by adding them to (when negative) or subtracting them from (when positive) annual operation-related costs.

LCC Calculation Methodology

Input information needed for LCCA includes cost estimates, by year, for two or more competing alternatives; a discount rate; and a study period. To calculate an LCC the present value of each cost to be incurred during the study period must first be computed using appropriately chosen discount rates. These present values for each alternative must then be summed, yielding its LCC. If other performance features are similar among the alternatives being considered, the alternative with the lowest LCC is the most cost-effective of the P2 investment options.

Calculations can be performed manually for simple systems, with the use of computer-based models for more complex systems and integrated systems. The following is the general formula for the LCC present-value analysis:

$$LCC = \sum_{t=0}^{N} C_t/(1+d)^t, \qquad (1.4)$$

where

LCC = total LCC in present-value dollars of a given alternative;

C_t = the sum of all relevant costs, including initial and future costs, less any positive cash flows, occurring in year t;

N = the number of years reflecting the life of the investment;

d = the discount rate used to adjust cash flows to present value.

The formula requires that all costs be identified by year and amount. Although the formula looks simple, the actual calculations can be quite tedious, especially when the life expectancy of the investment is more than a few years long, and for annually recurring amounts for which future costs must first be calculated to include changes in price.

The LCC method provides a consistent means of accounting for all costs related to several potential P2 investments over a specified study period. The LCCA is needed to demonstrate that the additional cost for an investment alternative is more than offset by its corresponding reduction in operating and maintenance costs (including energy and water), relative to the current situation. LCCA allows for a relatively straightforward comparison between tier 1 and tier 2 costs for P2 and pollution control technologies but often requires additional analysis to properly account for tier 3 and tier 4 costs. This additional analysis is referred to as *supplementary measures*, which are discussed next.

When using the LCC method for evaluating potential P2 investments a company should

- Try to choose among two or more mutually exclusive alternatives on the basis of lowest LCC.
- Design all alternatives to meet specific minimum performance requirements.
- Evaluate all alternatives using the same starting date, service date, study period, and discount rate.
- Subtract positive cash flows (if any) from costs.
- Ensure that any effects of a P2 investment not accounted for in money terms are either insignificant or uniform across all investment options or are accounted for in some other way.

Using Supplementary Measures of Economic Performance

Supplementary measures most often used to complement LCCA are

- Net savings (NS) measure.
- Savings-to-investment ratio (SIR).
- Adjusted internal rate of return (AIRR).

- Discounted payback (DPB).
- Simple payback (SPB).

Each of these supplementary measures is a relative measure of economic performance. They are most appropriately computed for a potential investment relative to the current situation. For assessing P2 investments, the "current situation" is most often a pollution control technology. The pollution control technology usually has lower investment costs (because it is already in place) and higher operational costs than the alternatives being considered. The main reason for LCCA of a P2 alternative is to demonstrate that its operational savings are sufficient to justify its additional investment cost.

The net benefits measure of economic performance is the difference between present-value benefits and present-value costs for an investment over its useful life. The NB measure is used in situations where positive cash flows are intended to justify an investment. A P2 example for this would be investment in a technology for producing biosolids for soil conditioning and agricultural markets from municipal wastewater sludge.

The net savings measure is a variation of the NB measure. The NS measure is used when benefits occur mainly in the form of future operational cost reductions, such as energy or water or chemical feedstock cost savings. The NS measure calculates the net amount that a P2 investment is expected to save over its life, in present-value dollars. Since the calculation is made in present-value terms, the savings computed represent savings above the amount that would have been earned from investing the same money at the minimum acceptable rate of return (the discount rate). The NS measure for a P2 investment relative to the current situation is calculated by subtracting the LCC of the P2 alternative from the LCC of the current situation, the base case (BC).

$$NS = LCC_{BC} - LCC_{P2} \tag{1.5}$$

As long as NS > 0, the investment is considered to be cost-effective relative to the current situation. Hence, the requirement for a financially attractive P2 investment is that its LCC be lower than the LCC of the current situation. For multiple, mutually exclusive P2 alternatives, the alternative with the greatest NS measure is the same alternative that has the lowest LCC. As such, LCCs and NS measures are consistent with each other and the analyses are interchangeable. One advantage of LCC over an NS measure when evaluating several alternatives is that the former does not require that the "current situation" be brought directly into the calculations.

The NS measure can also be calculated from individual cost differences between the current situation and a P2 alternative. In other words, the differences between initial investment costs, energy costs, OM&R costs, and other costs can be used as the basis for the NS calculation. Although this requires more calculations than the simplified method already presented, we prefer the latter approach because costs within the four tiers can be tracked in the computations. Further, the same calculations are required to compute the SIR and the AIRR. These economic performance measures should be calculated along with the NS measure to assess P2 investment options more thoroughly.

The NS calculation based on individual cost differences can be computed from the following equation:

$$\text{NS}_{A:BC} = \sum_{t=0}^{N} S_t/(1+d)^t - \sum_{t=0}^{N} \Delta I_t/(1+d)^t, \qquad (1.6)$$

where

$\text{NS}_{A:BC}$ = NS in present value dollars of alternative A relative to the base case (BC);

S_t = the savings in year t in operational costs associated with the P2 alternative;

ΔI_t = any additional investment-related costs in year t associated with the alternative;

t = the year of occurrence (0 would be the base date);

d = the discount rate;

N = the number of years in the evaluation period.

The SIR expresses the relationship between the cost savings resulting from a P2 investment and its incremental investment cost (in present-value terms) as a ratio. The SIR is a variation of the benefit-cost ratio (BCR) described in *Green Profit* and elsewhere. In fact, we recommend computing both. The SIR and BCR are used when benefits occur primarily as reductions in operation-related costs. Both are relative measures of performance that can be calculated only in relation to the current situation or an alternative investment, a "base case." The same base date, study period, and discount rate must be used both for the base case and the P2 alternative. A P2 alternative is considered economically justified relative to the current situation when the SIR > 1.0. This means that savings as a result of the P2 investment exceed its incremental investment costs and its net savings are greater than zero. The SIR is useful only as a means of ranking projects relative to other independent projects. The SIR should never be used as a basis for choosing among mutually exclusive investment alternatives, but rather strictly as a guide for allocating limited investment funds.

The general formula for the SIR is

$$\text{SIR}_{A:BC} = \frac{\sum_{t=0}^{N} S_t / (1+d)^t}{\sum_{t=0}^{N} \Delta I_t / (1+d)^t}, \quad (1.7)$$

where $\text{SIR}_{A:BC}$ is the ratio of PV savings to additional PV investment costs of the (mutually exclusive) P2 alternative relative to the base case, and all other parameters are defined as before.

The AIRR requires that the SIR for an investment (relative to the base case) be calculated first. The following formula can be used:

$$\text{AIRR} = (1 + r)\, \text{SIR}^{N-1}, \quad (1.8)$$

where r is the reinvestment rate (expressed as percent).

The AIRR is an economic measure of performance expressed in terms of the annual rate of return on an investment. A single investment alternative is cost-effective relative to the base case when the AIRR exceeds the appropriate discount rate. When computing the AIRR of an alternative P2 investment relative to its base case (the current situation or an alternative investment), the same study period and discount rate must be used. As with the SIR, the AIRR can be used to evaluate a single P2 investment relative to a base case and may be used to rank independent projects when allocating a limited budget. Effects that are not measured in dollars are not included in the analysis and must be accounted for in some other manner.

The most common supplementary economic measure of performance is the payback, of which there are two calculation methods. Both methods focus on the time required to recover initial investment costs.

The discounted payback calculation requires that cash flows occurring each year be discounted to present value before accumulating them as savings and costs. If the DPB is less than the length of the service period, the investment is generally cost-effective. This is consistent with the requirement that the LCC of the P2 investment be lower than the LCC of the base case. An alternative payback criterion is more commonly used (i.e., the number of years allowed for the payback to occur) by choosing a time period that is considerably shorter than the project's intended service period. A drawback with this calculation is that capital replacement costs or increased OM&R costs can occur after the year of payback, and this could cancel out the apparent cost-effectiveness of the investment.

The simple payback does not rely on discounted cash flows in the calculation. The SPB also ignores any changes in prices, such as energy, water, and labor escalation prices during the payback period. As with the DPB, the acceptable SPB for an investment is set at an arbitrary time period that is considerably less than the intended service period. Calculations based on the SPB generally result in a shorter year of payback since undiscounted cash flows are greater than their discounted counterparts (this assumes a positive discount rate).

The general formula for a payback calculation is

$$\sum \left[(S_t - \Delta I_t)/(I+d)^t \right] \geq \Delta I_0, \tag{1.9}$$

where

y = the minimum length of time (in years) over which future net cash flows have to be accumulated to offset initial investment costs;

S_t = the savings in operational costs in year t associated with a given alternative;

ΔI_0 = the initial investment costs;

ΔI_t = the additional investment-related costs in year t, other than initial investment costs;

d = the discount rate.

Between the DPB and SPB calculation, the former is preferred. However, both payback measures ignore all costs and savings and residual values occurring after the payback date. Payback calculations should be used only as a basis for screening potential P2 investments. The value in these calculations is their ability to help identify single P2 investments that are so clearly economical that the time, effort, and expense of a full LCC are not warranted. If there is uncertainty about the useful life of an investment, then the DPB method can also be used to determine an acceptable lower bound on its usefulness. Finally, it is important to note that a payback calculation is not a proper basis for selecting among alternative mutually exclusive P2 investments. Only the LCC and NS methods should be used for this purpose. In addition, payback methods should not be used for ranking potential investments for funding allocation purposes.

Liabilities for Simply Meeting Minimum Compliance Requirements
Introducing the Concepts
While there are exceptions, it is common among companies to heavily discount and even ignore those costs for compliance associated with intangible elements leading to future liabilities because of a false sense of security derived from meeting the minimum current

regulatory obligations. When a company looks at its operations only from a compliance standpoint, there is a tendency to focus on current regulations with perhaps no more than a few years of foresight into how regulations may become more stringent over time and their subsequent impact on the business. While no one can predict the changing regulatory climate, history has shown that, over the long run and worldwide, there has been a tightening of regulations across all forms of pollution media. Clearly, companies are in a position to assess their potential liabilities and, if nothing else, determine with some degree of confidence that their business operations pose certain risks and those risks can result in financial liabilities at some point in time.

Future liabilities could be sizable, and large companies that are "deep pockets" and sometimes big targets for the press can be especially vulnerable. Future liabilities associated with large remedial actions from past environmental damages run the risks of incurring large negative impacts from the "less tangible" cost tier, such as a negative public image and loss of investor confidence.

But, even when the more astute business decision makers realize the risks, there are various reasons or, more appropriately noted, excuses for discounting future and less tangible costs:

1. "We don't have a crystal ball and can't predict with reasonable accuracy how regulations may change or become more restrictive in the years to come."
2. "No formal guidelines for unsafe exposures currently exist for the chemical wastes we generate. We therefore don't believe the added costs for segregation, stabilization, on-site treatment, or off-site disposal to a secure location are justified."
3. "We meet our regulatory obligations. Everything we do is within what is allowed by the state in which we operate our business. Since we are following the law, we have no exposure."
4. "We convinced our state environmental regulators that our current waste management practices are adequate and they are issuing us permits. We have nothing to worry about now and in the future."

5. "We don't need to worry about any future liabilities because we engaged a company that will manage our wastes. But, in addition, we cleverly structured an agreement whereby they accept 100% of all the liabilities associated with the wastes we have contracted them to remove from our property and dispose of."

On the surface, nothing seems to be wrong with these statements. The following case studies, however, reveal fatal flaws in the environmental management practices of several well-known large corporations. These flaws have led to tens of millions of dollars in legal fees over disputes that are in the league of hundreds of millions of dollars in settlement damages. The case studies described are taken from litigations in which I have been involved as an expert witness over the past decade. Since these are litigations in which testimonies have already been given, they are a matter of public knowledge, and testimonies and evidentiary literature are available in the public domain. Consequently, no interviews or comments from representatives of the respective companies cited here were obtained in preparing this discussion. Indeed, it would be surprising if any of these companies would want to discuss the issues openly in light of the magnitude and seriousness of the litigations, which involve toxic torts with large numbers of communities. For those readers not familiar with the terminology, *toxic tort* refers to litigation concerning personal injury caused by exposure to a toxic substance, such as asbestos or a hazardous waste. Victims can sue for medical expenses, lost wages, and pain and suffering.

While various experts expressed opinions criticizing each corporate defendant for irresponsible actions over different time periods, the intent of the following discussions is not to comment on whether these companies acted irresponsibly or even acted innocently when their actions resulted in negative environmental impacts. Rather, the examples point toward the need to recognize and consider the risks of manufacturing operations, their potential future exposures to a company, and the absolute necessity to manage environmental aspects in an aggressive and proactive manner.

Case Study 1. Redlands versus Lockheed-Martin

"We don't have a crystal ball and can't predict with reasonable accuracy how regulations may change or become more restrictive in the years to come."

"No formal guidelines for unsafe exposures currently exist for the chemical wastes we generate. We therefore don't believe the added costs for segregation, stabilization, on-site treatment, or off-site disposal to a secure location are justified."

Lockheed-Martin is a large defense contractor that manufactured rocket propellant motors for the U.S. Department of Defense at a plant in Redlands, California. Between 1954 and 1976, it manufactured rocket motors in fulfillment of contracts for a variety of weapons and satellite systems, ranging from air-to-surface missiles, antitank missiles, and space exploration vehicles.

Lockheed used a dry propellant technology that consisted of mixing rocket fuel formulations in a wide range of rubber matrix systems. The principle ingredient used was ammonium perchlorate (about 65% or more of the rocket fuel mix consisted of ammonium perchlorate, or AP).

The Redlands facility where Lockheed operated its plant was a land area that sat over a recharge basin for a drinking water aquifer. Lockheed selected the site for its operations in 1954 because it had good infrastructure (e.g., water supply, electricity/power, access roads) and was in close proximity to a workforce.

This was a large manufacturing facility with a workforce exceeding 1,000 at times. Because it was a government contractor, it was required to produce its products to very stringent specifications that required tight tolerances and high performance. To meet product performance specifications, considerable pilot testing was performed over the years. Pilot testing consisted of developing new rocket propellant formulation mixes and experiments aimed at integrity and failure mode analyses. This included test firings on the property.

To save money, metal motor casings were reclaimed after test firings and recycled for use. The residues from test firings became a part of the solid waste streams from the plant.

Rocket motors were composed of metal casings into which the AP-rubber matrix was oven cured. The propellant was manufactured in a variety of ways, ranging from conventional compounding operations using mills and large capacity mixers, in which the propellant or rocket fuel was poured into the casing and cured in large bunker-type ovens, to extruders that relied on injection molding of the propellant into the metal casing.

The casing consisted of metal parts with metal-to-metal seals. All metal parts required meticulous cleaning prior to matching seals and molding operations. Metal parts cleaning was performed with vapor degreasers, which used thrichloroethylene (TCE), which is a chlorinated hydrocarbon and a suspect carcinogen. TCE was extensively used at the site for other purposes. Other purposes included tool cleaning and the cleaning of machines (mixers, blenders, extruders that were used for rocket fuel compounding purposes).

While the RCRA (Resource Conservation and Recovery Act) did not exist during the time of the operations, the facility would have qualified as a large-quantity generator. The solid wastes that were burned on-site ran into the millions of pounds with waste burning occurring five to seven days a week and at multiple times of day, depending on production levels.

Lockheed managed the wastes generated at the site with simple, nonpassive systems that consisted of concrete-lined evaporation ponds and surface impoundments. But it also performed open-pit burning of wastes on bare ground. Much of this waste was spent AP from mixer cleanout operations and the residues from test fire motor casings.

Lockheed burned many millions of pounds of AP waste on their site. In addition, it had horrendous housekeeping problems, ranging from spent solvents (TCE, in particular) being thrown away in an uncontrolled manner by employees at times to the washing of building floors and processing equipment, where liquid waste streams were sent to pits, sumps, and sinkholes. While Lockheed maintained a large number of evaporation ponds and surface impoundments to contain much of its wastes, evidence suggests that these pits overflowed at times during period of excess production.

In the middle to late 1990s, both AP and TCE were detected in groundwater. More than 1,000 individuals were exposed to contaminated drinking water directly attributable to the Redlands facility operations. This led to a major class action lawsuit against Lockheed for health damages.

Lockheed argued that, during the historical period of the plant's life, neither AP or TCE or many other chemicals managed at the facility were regulated as hazardous wastes. The company further argued that the waste management practices that Lockheed relied on, while considered illegal and totally unacceptable by today's standards, were no different from other facilities in various industry sectors for that time period. They also argued that there were few if any health risk assessments on AP and the maximum allowable limits in groundwater and drinking water did not become legal standards until the 1990s.

While all these are factual, Lockheed still had an obligation to act in a responsible manner. While Lockheed had no pressure to protect the groundwater from a regulatory standpoint, the facility knowingly operated over a recharge basin that was established as a drinking water source from as early as the 1920s. Lockheed was irresponsible in having selected the site for its operations to begin with, in 1954.

The plant's location and the nature of its operations mandated the best available technologies for waste management. Simply because other industries and companies acted irresponsibly is no excuse for actions that might border negligence. Lockheed could have installed wastewater treatment technologies to manage liquid wastes containing AP and TCE. Instead, it chose the lowest-cost technology of the day. By the late 1950s, the concepts of fate and transport in groundwater flows were highly developed and understood. And, indeed, throughout the 1960s, there was a keen awareness and thrust against Congress to enact stringent environmental legislation. While Lockheed argued that not much toxicological data existed on AP at the time of its operations, it does not take a rocket scientist to figure out that it would not be good to allow rocket fuel to enter into a drinking water aquifer. Placing the facetious remarks aside, significant literature existed at the time on the toxicity of TCE and the more than 200 other chemicals managed at the facility.

There were no lack of treatment technologies for wastewater; other companies were using on-site incineration technologies to manage solid wastes by the late 1960s; some companies were recycling and conducting source reduction and waste minimization programs. A review of records spanning nearly a quarter of a century on the facility operations did not show any of these activities were adopted by Lockheed-Martin.

In my opinion, Lockheed acted irresponsibly in protecting the public from its wastes. Whether or not that was the norm by industry on the whole is immaterial. Redlands is a unique situation because it is likely the only site in the United States where a chemical manufacturing operation of this magnitude was operated on a drinking water recharge basis.

The engineering literature shows that site selection criteria aimed at protecting waterways and groundwater existed as early as the 1920s. Toxicology as a science began in the United States in the 1930s and, by the 1950s, it was commonplace for chemical suppliers to provide documents that were the precursors to modern MSDS (material safety data sheets). This means that Lockheed had technical information on the hazardous and toxic nature of the chemicals it managed at the site. This further means that it had a fundamental understanding that, if wastes containing those chemicals reached the groundwater aquifer, they created a pollution problem.

Lockheed remains entangled in a large toxic tort because of the company's actions from decades ago. Like other companies of that era, Lockheed lacked the foresight to predict the costly litigation of a toxic tort and the massive remediation effort it faced, but it should have understood at the time that the operations created a potential risk to the public.

This brings us to an important point. While the concept of an EMS did not exist during the time period of the Redlands site operations, the concept of acting in a responsible manner with regard to protecting the public and the environment did. The societal expectations of the day that industry act responsibly and take precautions against public exposure to hazardous chemicals was in itself a standard to which Lockheed and others from that era were expected to follow. Furthermore, trade organizations and industry associations like the API (American Petroleum Institute) and MCA (Manufacturing Chemists Association) published voluntary guidelines for waste treatment technologies and best management practices. Because of the proximity and nature of its operations, Lockheed should have applied the best available technologies of the day. It should have gone beyond the general norm of poor to arguably average practices that other industries and companies followed.

Some readers may be skeptical and argue that Redlands is a poor example because U.S. companies do not act this way today. In a modern context, the skeptical reader may argue that poor housekeeping practices no longer exist and companies are highly diligent in eliminating the obvious waste problems. They may further argue that everyone today maintains a waste minimization program of some sort: hence, the kind of health risks and exposures created by gross mismanagement of wastes does not happen anymore. To those critics, I say read on.

Case Study 2. Tie Plant Litigation

"We meet our regulatory obligations. Everything we do is within what is allowed by the state in which we operate our business. Since we are following the law, we have no exposure."

"We convinced our state environmental regulators that our current waste management practices are adequate and they are issuing us permits. We have nothing to worry about now and in the future."

The Tie Plant wood treatment facility located in Grenada, Mississippi, is owned and operated by Koppers Inc./Beazer Corp. The facility has been in almost continuous operation since 1904. Creosote and its solutions in petroleum oils and pentachlorophenol solutions in hydrocarbon oils were and continue to be used as treatment chemicals. The company is entangled in a multimillion dollar toxic tort instigated by a neighboring residential community.

From its inception, the Tie Plant facility site contained, among its operating units, cylindrical treatment retorts, storage tanks, work tanks, heating units, boilers, cooling towers and condensers, waste-liquid pits, tie and pole storage yards, ditches used for waste discharges, waste-liquid and sludge ponds, woodworking and solid waste disposal areas. These operating units generated waste fluids continuously and in significant volumes. Until at least the late 1980s, none of the facility's operating units were equipped with pollution control devices that would have reduced releases of contaminants.

Tie Plant wood treatment operations routinely generated large quantities of waste fluids as chemical liquids. The physical forms of such wastes were manifested as wastewater containing dissolved and suspended chemical constituents, aerosol droplets containing dissolved constituents, vapors of volatilized chemical liquids, particulates containing chemicals, spent treatment chemicals as free-phase liquids, and hazardous solids contained in sludge wastes.

For more than seven continuous decades, the standard practice of waste disposal was uncontrolled discharges into surface water bodies through drainage ditches, onto on-site surface soils, into unlined impoundments, and into the local atmosphere. Such discharges were in uncontrolled communication with surface water flows and groundwater flows through the facility site to down gradient off-site locations. A review of corporate documents revealed that the total plant effluent in 1969 was 50,000 gpd. Some documents reported 1.8 million gallons per month (21 million gallons per annum) as the total plant effluent. The defendant allowed these volumes of toxic wastewaters to overflow limited-capacity earthen ponds into drainageways leading to surface watercourses and to drain onto off-site land areas where a neighboring community resided. The owner/operator of Tie Plant, along with a codefendant (Illinois Central Railroad, or ICR) are entangled in a costly toxic tort. ICR's involvement is explained in case study 3.

This complex litigation can be reduced somewhat by examining different periods of historical operations: those operations that lead to wastes and the management practices applied prior to the advent of strict environmental enforcement, and the period where environmental regulations and enforcement came into effect. The following comments are based on the latter period.

The facility continues to operate despite excessive levels of on-site contamination from creosote, pentachlorophenol, and dioxins. Rather than overwhelm the reader with discussions on the many environmental aspects associated with this 171-acre facility, attention is given to one environmental aspect, which is a wood-waste boiler.

Circa 1979, the facility installed a wood-waste fired boiler that burned untreated wood. As with most wood treating facilities, considerable scrap is generated that can be wisely converted to steam and energy to run the plant's treating cylinders. Tie Plant also generates large volumes of toxic wastes in the form of wastewaters, sludges, and scrap treated wood. The facility also cumulatively generates large amounts of contaminated soil from drippage and spillage of treating chemicals. Disposal of such wastes is costly and ongoing, especially if transported and disposed of off-site at secured landfills and by licensed waste treatment facilities.

The company came up with what seemed to be two clever and profitable concepts, both of which were initially sanctioned by the Mississippi Department of Environmental Quality (MDEQ). The first concept was to burn the toxic sludge in the boiler. Limited pilot tests were performed circa 1984 in which the operator demonstrated that dioxin emissions generated during combustion underwent 99.99% DRE (destructive removal efficiency). While the emissions tests supported that no significant discharges of dioxins were generated, the tests were brief and in part based on tests conducted at another similar but not identical boiler. To add further justification, the owner lobbied for a permit based on the following arguments presented to the MDEQ:

1. Process wastes generated from the plant were not regulated hazardous wastes. This was a matter of the state of Mississippi's interpretation of waste definitions under the RCRA. The consequence of this interpretation was that the proposed action was not subject to regulatory enforcement.

2. The program was presented as a form of waste recycling, whereby the owner/operator argued that the plant efficiently

converted waste to energy, thus implementing a green technology or rather a cleaner production program.

Both these arguments and the limited combustion tests formed the basis for an air pollution permit that enabled toxic sludge to be burned in the on-site wood-waste boiler. The permit limited the waste to a 5% fuel additive program (i.e., no more than 5% of the total fuel could be composed of process waste, and the furnace temperature had to be maintained at temperatures of around 1800°F to ensure a high DRE of dioxins, furans, and PAHs (polycyclic aromatic hydrocarbons), which are known to be generated from the combustion of creosote- and pentachlorophenol-derived wastes. The permit did not call for any substantial modifications to the boiler and the only air pollution control used was a multiclone with a particulate removal efficiency of between 85% and 90% particulates.

Koppers/Beazer argued that it had a permit and operated within the legal requirements of the state of Mississippi. On the surface, the company created a legitimate program that could be argued is cleaner production, enabling wastes to be responsibly managed and converting this waste to a useful form, namely, energy for on-site consumption. On the surface, this program had financial and environmental credits associated with it. Financial credits were derived from a reduced dependency on supplemental oil and natural gas during cofiring operations of the boiler and the elimination of waste transport and disposal at licensed off-site treatment and disposal facilities. From the standpoint of waste management, wastes would never leave the generator's site and be effectively managed and even destroyed or immobilized at the source. On the surface, this seemed like an attractive program that embodies sound environmental management.

As with many litigations evolving around historical industrial practices, records are often incomplete. However, surviving records showed that the owner/operator burned 5,980,821 pounds of sludge over a 35-month period from 1988 to 1990. Based on average monthly incineration rates and knowing the life of the practice, it was determined that the facility more likely than not burned as much as 19.3 million pounds of sludge between 1982 and June 1992. The incinerated sludge was contaminated with high levels of heavy metals (including, without limitation, arsenic, chromium, copper, and zinc), PAHs, and phenols. These wastes should never have been allowed to be burned in a boiler that did not undergo substantial modifications, especially since it had not been demonstrated that the boiler could maintain high temperatures on a consistent basis.

Where the owner/operator acted most irresponsibly was as follows. The company accepted process sludge from at least 17 other facilities, including those derived from Superfund sites and facilities not owned or operated by them. In essence, it used its permit status to circumvent the intent of the law, to protect the public and the environment. While the sludge wastes from other facilities located in other states were considered RCRA-regulated hazardous wastes, the owner/operator took advantage of the liberal waste classification interpretation of the MDEQ to enter into the waste disposal business. In short, it had wastes shipped to the Tie Plant where it incinerated the wastes for profit through its fuels additive program.

A review of the company's internal records showed that it exceeded federal air pollution guidelines at times by burning sludge with excess levels of heavy metals and at burn rates that exceeded its air pollution control permit and the pollution control limits of the equipment. It burned toxic sludge on a near continuous basis: on average, it burned large quantities 24 days per month; on average, it burned large quantities of sludge 16 hours per day (and many times up to 24 hours per day on a continuous basis); it burned as much as 15,000 pounds of highly toxic wastes during a single daily burn period.

The owner/operator used a wood fired boiler without recognized air pollution controls to incinerate hazardous industrial wastes. In fact, earlier internal memos from the 1970s show that the corporation acknowledged that the incineration of sludge of the compositions burned in its wood-waste boiler required ultrahigh-combustion temperatures (above 1800°F) to minimize hazardous air emissions. The boiler was not designed to nor was capable of achieving temperatures on a consistent basis that would ensure complete destruction of the toxic components in the sludge materials burned.

The boiler was never retrofitted with catalytic afterburners and an ESP (electrostatic precipitator) to handle the high loadings managed at times. Furthermore, the boiler was used for combusting and roasting contaminated soils derived from spills. The boiler technology employed was never designed for that purpose.

Circa 1992, the MDEQ changed the status of the wastes incinerated by Koppers/Beazer from nonregulated process wastes to RCRA-regulated wastes. Since the boiler was not permitted for incinerating wastes under the classification, the company complied by terminating the fuels additive program.

Koppers/Beazer argues that the company operated entirely within the law. While records show it exceeded federal air pollution guidelines at times, overall the program was legally permitted and it complied with terminating the operation when the regulatory status for the wastes changed.

While Koppers/Beazer generally operated within the law; that is, it met the minimum legal compliance requirements by having a permit to operate, it acted irresponsibly. All environmental laws are understood to provide minimum standards aimed at protecting the public and the environment. Because legal standards are based on the premise of protection based on the overall good of society, they do not account for special circumstances. In the case of Tie Plant, the special circumstances are that this is a large wood treating facility immediately adjacent to a residential community. Its neighbors are sensitive receptors, hence there is the potential for negative impacts on humans from its operations. Further, while the MDEQ did not interpret the process sludge as a regulated hazardous waste in accordance with the RCRA definitions, all other states that Koppers/Beazer accepted wastes from did. The nonregulated waste classification interpretation on the part of the MDEQ was based on a legal definition and was not justified based on health-risk assessments or challenges levied against other states on the basis of waste classification being overly conservative by designating wood-treating–process sludge as regulated hazardous wastes. Koppers/Beazer simply used the law to its advantage to capitalize on an opportunity to manage wastes from its many facilities in a highly cost-effective manner and accept waste for a fee. The company, in essence, used its permit status to enter into the waste management business. It further deceived the MDEQ because it did not disclose to the state regulatory agency its intent to accept wastes from 17 other states and facilities.

Koppers/Beazer's fuel additive program added to an already serious waste management problem at the facility. For the few surviving burn records covering the period between 1988 and 1990, more than 11,000 drums of toxic waste sludge were handled. There are no records to show whether these drums, a hazardous waste stream unto themselves, were managed in an environmentally responsible manner. The history of the facility shows mismanagement of spills and poor housekeeping practices that contributed to site contamination for many decades. One therefore would not expect spent drums containing toxic residues to have been any more responsibly managed than other wastes at the facility.

Koppers/Beazer's fuels additive program did the exact opposite of what the MDEQ thought it would. Instead of streamlining waste management at Tie Plant, it expanded waste management activities by accepting nearly three times more wastes from other facilities for the period where records survived. The program created serious health risk exposures from environmental transformations to air pollution and ash containing heavy metals that were stockpiled on-site and became airborne. It added incrementally to global air pollution problems by allowing Tie Plant to act as a waste disposal facility. In other words, the potential environmental credits that could have been claimed by the elimination of off-site trucking and disposal of wastes were significantly offset by the transportation and on-site management of thousands of toxic waste drums.

This portion of the case study should demonstrate to most readers, if not all, that meeting minimum regulatory requirements does not make for a responsible corporate citizen. In fact Koppers/Beazer's actions might very well be described as reckless endangerment.

A Second Folly

The second clever and potentially profitable concept introduced by Koppers/Beazer also centered around the wood-waste boiler operations. In this case, the company petitioned the MDEQ to burn discarded treated wood in their wood-waste boiler.

On the surface, the concept proposed appeared to be environmentally friendly and a cleaner production-type program. In fact, it is practiced at several facilities around the United States but with safer technologies than employed at Tie Plant.

Koppers/Beazer proposed a program whereby it was paid by the Illinois Central Railroad to receive old treated railroad ties, which were burned in the wood-waste boiler to generate steam and electricity. On the surface, this waste-to-energy program offered the environmental credit of reducing landfill dependency and converting a waste into power that could be both used on-site and sold to a local power grid.

Once again, the program had the opposite impact. While waste volume was reduced by over 90%, thereby reducing the dependency on landfill disposal practices, these credits were offset substantially by the air pollution created from incineration. Permits were simply obtained without full demonstration of the reliability of the boilers' limited air pollution controls and required high thermal destruction ability. Throughout the history of this operation (from circa middle to late 1990s through 2001), the owner/operator's records show that the boiler could not consistently maintain sufficiently high combustion temperatures to ensure an adequate level of DRE of the dioxins and furans generated as products of incomplete combustion of treated wood. This, in part, was related to inconsistent feeding problems to the boiler, which could never quite be optimized due to hardware limitations. Again the owner/operator ignored the special circumstances these operations impose on the quality of life and health of sensitive human receptors. The capacity of the boiler operations is such that they emit more than 60 tons per year of particulates and air emissions from the boiler alone. The owner of the facility added enormous health exposure risks to the community with exposure to dioxins and furans by implementing this practice. No substantial modifications were made to the boiler. Continuous stack monitoring was limited to CO and opacity, with crude temperature control and temperature sensors that were not reliable. Test fire data used as the basis to petition for the air permit for this practice was limited and largely based on tests from similar but not identical boiler operations.

The owner/operator appears to have met the minimum compliance requirements with this practice. It had a permit to operate. There are no records to substantiate that the company violated that permit either intentionally or unintentionally. Its exposure to a toxic tort in this case is substantial, with serious health claims that fingerprinted daughter compounds of dioxins and furans to pentachlorophenol and creosote from the plant.

Summary and Lessons Learned
Clearly, just operating within the legal requirements of environmental statutes does not ensure that future liabilities will not be incurred. The less tangible costs that are likely subsequent impacts from this litigation surfaced from the owner/operator's environmental insurance but also strong arguments arose that the site itself may be so badly contaminated that it qualifies for the Comprehensive Environmental Response, Compensation, and Liability Act (CERCLA) listing.

To date, Tie Plant has no formal EMS. It manages its environmental aspects the same way it always has—with a focus on compliance only. While numerous violations exist over the life of this facility during the era of regulatory enforcement, on the whole, this is a company that has shown diligence in attempting to meet compliance schedules on time. For the most part, it obtained the proper permits and maintains them. But meeting its minimum legal requirements did not prevent a community of human receptors from becoming terminally ill from air pollution exposures associated with the plant's operations. And being within compliance did not protect the company from very significant financial exposures from a toxic tort.

An EMS would have helped this facility focus on waste minimization, pollution prevention, and directed management's attention to proper planning for pollution control infrastructure investments that could have limited, if not eliminated, health risk exposures by the community and the financial liabilities for the company. An EMS would have drawn attention to the major environmental aspects associated with the manufacturing operations and enabled this company to recognize the need to address the aspects taking into consideration public safety and many of the elements distributed among the four cost tiers.

This case, more than others I encountered, demonstrates that an EMS is a matter of acting responsibly. It is a basis by which modern companies exercise responsible care in the safe management of the environmental aspects of their operations.

Case Study 3. The Role of Illinois Central Railroad in the Case against Koppers/Beazer

"We don't need to worry about any future liabilities because we engaged a company that will manage our wastes. But, in addition, we cleverly structured an agreement whereby they accept 100% of all the liabilities associated with the wastes we have contracted them to remove from our property and dispose of."

Illinois Central Railroad was a codefendant in the Tie Plant toxic tort, because ICR's old railroad ties were burned as waste in the boiler. This company has argued that it has been innocently dragged into the litigation because it had no on-site management of wastes, it entered into a good-faith agreement with Koppers/Beazer to dispose of its wastes, and the only reason it is entangled in the dispute is because ICR has "deep pockets."

ICR acted at times more than a mere customer with occasional on-site inspectors for QC (quality control) purposes. Internal corporate documents show that railroad personnel instructed Tie Plant operators at times to alter normal process operating conditions that influenced greater levels of waste generation. Aside from this issue, ICR's business practices resulted in solid wastes (old railroad ties) that were not managed responsibly.

Responsible companies manage their solid wastes from cradle to grave in an environmentally responsible manner, using best practices and technologies. This is known as *exercising environmental stewardship*. Responsible companies that act with a focus on protecting the environment and public safety show accountability for the products, by-products, and wastes created by their business practices. They direct attention, effort, and investments to not only comply with environmental regulations but take full responsibility for managing the resources and by-products, including wastes, in a manner that does not negatively influence the quality of life and the environment. The following are Web sites for companies that have formal corporate environmental policy statements that reflect these practices:

- Hormel Food Corp.—http://www.hormel.com/templates/corporate.asp?catitemid=71&id=314
- Walt Disney Corporation—http://www.corporate.disney.go.com/environmentality/environmental_policy.html
- 3M Corp.—http://www.solutions.3m.com/wps/portal/!ut/p/kcxml/04_Sj9SPykssy0xPLMnMz0vM0Y_QjzKLN4j3MwHJgFjGpvqRqCKO6AK-QRAR32CYiKU7uho_YwxdIVARA7hVjq4QIWNHX4SYO4aYtwVEyNtS39cjPzdVPyg1L97ZUd9bP0C_IDcUCiLKHR0VFQEHJRvR/delta/base64xml/L0lJYVEvd05NQUFzQURzQUVBLzRJVUZDQSEhLzZfMF8xMkwvZW5fVVMvMC8xMC82/
- General Motors Corp.—http://www.gmcanada.com/inm/gmcanada/english/about/Environment/environment.html
- CSX—http://www.csx.com/?fuseaction=general.csxo_env.

ICR elected to dispose of millions of pounds of a solid waste by relying on a specific technology applied at Tie Plant that created harmful environmental transformations. Records, in the form of contracts and correspondence between the two companies, show that ICR chose to dispose of its wastes (old treated railroad ties) with consideration given only to low cost and not as to whether the incineration practices relied upon by Tie Plant for disposal were safe and did not create air pollution problems and contaminated storm-water runoff. ICR attempted to pass on its ownership of the wastes its business generated by having the owner of Tie Plant enter into a sales agreement for the purchase of these wastes, thereby attempting to relinquish its obligation and making waste management another company's concern.

This is contrary to how businesses manage wastes in an off-site manner. Companies that generate wastes identify and enter into agreements with waste management and disposal facilities that have established expertise and rely on technologies based on the best available ones. But, even in those agreements, the original companies do not relinquish ownership of their wastes. Under U.S. environmental legislation, and indeed the same premise applies throughout the European Community, a waste generator always maintains liability for the wastes it generates. For this reason, a generator must always exercise a high degree of due diligence in selecting third parties to responsibly manage its wastes.

ICR knew that Tie Plant operated a wood-waste boiler that, while eventually permitted, was not designed for waste incineration. Correspondences between the two companies shows that ICR was kept informed of licensing status and that the licensing was not based on substantial modifications to the boiler. ICR understood the hazardous nature of its solid wastes, especially when that waste is combusted, creating dioxins and furans. Like any other company that generates a waste product, it knew the properties and toxic characteristics of its own wastes. Like any other company, it has an obligation to select a waste management company that relies on technology that ensures dioxins are not generated as hazardous air pollutants.

Koppers/Beazer was not an established waste processing business, and it had no prior expertise or experience incinerating wastes that generate hazardous products from incomplete combustion. A company that acts responsibly would have further inquired and assessed whether the hardware used to manage its solid wastes was appropriate. The records show that, while Tie Plant had relationships with other customers that generated treated wood wastes, only ICR wastes were burned. And, as the Tie Plant's records show, the practice of incinerating treated wood wastes was terminated because the owner/operator eventually admitted that the boiler could not be operated in a safe and consistent manner and the company's liabilities were ever increasing from the practice.

ICR has an obligation to manage wastes from cradle to grave. Like all other companies, it is accountable for its wastes. No company, except in third-world nations, simply signs off its obligations for managing their wastes to a third party and ignores whether or not those wastes are being responsibly managed.

ICR manages a large business with many environmental aspects, including but not limited to warehouses for parts, chemicals, and fuel; properties that maintain storage tanks; engine and parts cleaning and maintenance operations using solvents operated at its sites, where there are hazardous wastes derived from battery waste and sludge from wastewater treatment; and operations that create air emissions.

Companies that manage operations with many environmental aspects need an EMS. An EMS helps collapse the many issues within a single management framework that enables senior management to focus attention on the highest priorities and evaluate risks to the business. It helps to direct a company to act responsibly for all of its environmental issues, not only within the sites it operates but along the entire supply chain.

The Difference between an Environmental Program and an EMS

An environmental management system (EMS) is a systematic approach to handling the environmental aspects of an organization. It is a management tool that enables an organization of any size or type to control the impact of its activities, products, or services on the natural environment.

Every company today will claim that it has an environmental management system. It will point to the fact that it has an environmental manager or that the plant manager is responsible for environmental reporting issues, permits, meeting compliance requirements, and the like. It will point to the fact that it has personnel and perhaps even a dedicated department that manages the environmental aspects of its business operations. Some will highlight the fact that they have a formal waste minimization program. However, this is not an EMS but a collection of components loosely linked together and applied to address environmental compliance issues.

There is a fundamental difference between managing environmental issues with corporate resources under a program within an environmental department and adopting an EMS, which is a systematic approach to linking and managing all aspects of a business to the environmental interactions of the operations and products.

The overall lesson from the preceding case studies is that none of the companies maintained an EMS. They face multimillion dollar litigations because they have not managed their environmental affairs in a proactive manner. Whether the claims by the plaintiffs are substantiated or not is immaterial. A court of law decides this. But, in the interim, each company faces costly and time-consuming legal battles.

As corporations, they reacted to environmental problems rather than addressing them proactively. Their approaches to environmental management are fundamentally wrong and, in fact, irresponsible, because all companies have an obligation to protect the public at large from their operations. An environmentally responsible company recognizes that it must continually improve its environmental performance by systematically addressing and eliminating the negative environmental aspects of their manufacturing operations.

Having the legally required documents, such as a storm-water management plan, a spill contingency plan, waste minimization reports, and air pollution permits, unto themselves, does not constitute proper management of the environmental impacts that a business imposes on the public and the environment. What a company does when it focuses only on the permits and regulatory mandates is meet its legal requirements. What it does not do is assess whether meeting the minimum requirements are adequate in terms of mitigating risks to the community and its own business. A clear indication of this is the fact that, in case studies 2 and 3, the companies could not even produce a complete set of production and waste records for the last decade of their operations. A company that cannot produce such records simply is not tracking its environmental performance and does not have a focus on continually improving that performance.

The point of all this is not to convince the reader that any of the companies discussed is irresponsible and even reckless. That is for a jury to decide, based on many other factors not discussed. The overall point is that each company brought a class action suit on itself by its own poor environmental management practices. The company relied on a system that in fact is not a system. It relied on a general approach to managing their environmental aspects. Herein lies the difference between managing environmental issues with a program versus the application of an EMS.

The key elements of an EMS are

- *An environmental policy.* The environmental policy and the requirements to pursue this policy via objectives, targets, and environmental programs.
- *Planning.* The analysis of the environmental aspects of the organization (including its processes, products, and services as well as the goods and services used by the organization).
- *Implementation and operation.* Implementation and organization of processes to control and improve operational activities that are critical from an environmental perspective (including both the products and services of an organization).
- *Checking and corrective action.* Checking and corrective action, including the monitoring, measurement, and recording of the characteristics and activities that can have a significant impact on the environment.
- *Management review.* Review of the EMS by the organization's top management to ensure its continuing suitability, adequacy, and effectiveness.
- *Continual improvement.* The concept of continual improvement is a key component of the environmental management system; it completes the cyclical process of plan, implement, check, review, and continually improve.

An EMS is based on a set of standards. There are several versions of an EMS, but in the United States, we recognize the ISO 14000 series. The standards in the ISO 14000 series fall into two major groups: organization-oriented standards and product-oriented standards.

The organization-oriented standards provide comprehensive guidance for establishing, maintaining, and evaluating an EMS. They are also concerned with other organizationwide environmental systems and functions.

The product-oriented standards are concerned with determining the environmental impacts of products and services over their life cycles and with environmental labels and declarations. These standards help a business gather information it needs to support its planning and decisions and communicate specific environmental information to consumers and other interested parties.

Understanding the Elements of an EMS

The Basics: The Continual Improvement Cycle

A simplified view of any EMS is in terms of the continual improvement cycle. In essence, what an EMS challenges an organization to do is continually improve its environmental performance. By continually improving the environmental performance of a facility, even when the improvements are incremental, over time, the performance improves measurably. The continual improvement cycle consist of four basic steps:

1. Plan.
2. Do.
3. Check.
4. Revise.

Let us look at Figure 1–1. In the planning stage,

- We formulate environmental policy.
- We identify environmental compliance needs.
- We identify and set priorities for other environmental performance improvement needs.

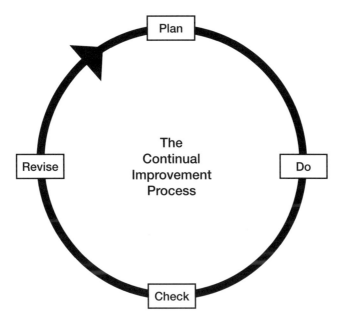

Figure 1-1 The continual cycle process.

- We design programs and action plans to improve environmental compliance and performance.

In the doing or implementing stage, we implement the environmental programs and action plans devised under step 1.

In the "check" step,

- We monitor environmental performance.
- We evaluate the progress of the environmental programs and action plans.

In the "revise" step,

- We revise the environmental programs and action plans.
- We consider changes in the environmental policy.
- We consider new environmental compliance and other performance improvement needs.

This is all common sense. In fact, these are the same steps decision makers follow when making business decisions. An EMS is simply a management system that follows a formalized set of steps that logically address a plan.

Now let us talk about ISO 14001. ISO stands for the International Organization for Standardization, which is a worldwide federation of national standards bodies, at present comprising 127 members, one in each country. The object of ISO is to promote the development of standardization and related activities in the world with a view to facilitating international exchange of goods and services and to developing cooperation in the spheres of intellectual, scientific, technological, and economic activity. The results of ISO technical work are published as International Standards.[3]

ISO 14000 is a series of international, voluntary environmental management standards. Developed under ISO Technical Committee 207, the 14000 series of standards address the following aspects of environmental management:

- Environmental management systems (EMS).
- Environmental auditing and related investigations (EA&RI).
- Environmental labels and declarations (EL&D).
- Environmental performance evaluation (EPE)
- Life-cycle assessment (LCA).
- Terms and definitions (T&D).

The ISO series of standards address the needs of organizations worldwide by providing a common framework for managing environmental issues.

ISO 14001 is the part of the standard most pertinent to our discussions. Figure 1–2 relates the five basic steps and elements essential to implementation of the continual improvement cycle. For a detailed discussion of each of the clauses contained in ISO 14001, the reader may refer to the publication *Green Profits*.[4]

---

[3]ISO Memento 1998 (Geneva: International Organization of Standards, 1998).
[4]N. P. Cheremisinoff and A. Bendavid-Val, *Green Profits: A Manager's Handbook to ISO 14001 and Pollution Prevention* (Oxford: Butterworth–Heinemann Publishers, 2001).

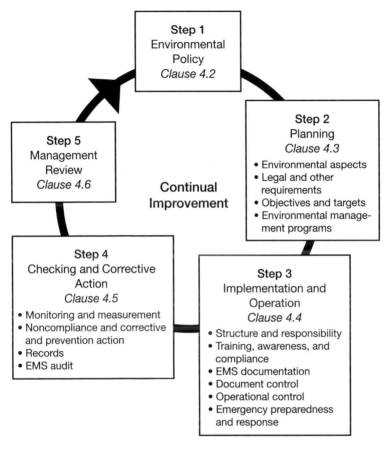

Figure 1–2 The main clauses of ISO 14001 that apply to the continual improvement cycle.

What Conformity Assessment Is All About

The term *conformity assessment* refers to a situation in which an organization tries to determine if the requirements of a standard are being met; that is, it is an audit. In the case of a management system standard such as the ISO 14001 EMS specification, conformity assessment by an accredited third party is the basis for the "certification" or "registration" of an organization to the standard.

Just as with the ISO 9000 quality standards, the value of an ISO 14001 certification depends on the confidence others have in the body that performs the certification and the process it uses. There must be assurance that the certification was performed rigorously and fairly. This confidence is provided through the process of accreditation, which embodies the recognition that a certifying body is qualified to do the work.

In the United States, the accrediting organization is the American National Standards Institute/Registrar Accreditation Board (ANSI/RAB). For registration (certification) to ISO 14001, a company first requests (and pays for) an audit by a RAB-accredited registrar (note that the registrar may be accredited by a counterpart organization in another country). The registrar then hires, fields, and supervises a team of RAB-certified ISO 14001 auditors, including a RAB-certified lead auditor. Auditors may be certified by a counterpart organization in another country. RAB-certified auditors must have successfully completed training by a RAB-accredited training provider and meet other requirements, including annual recertification requirements (auditing hours, professional development, etc.). Figure 1–3 provides a summary of the accreditation infrastructure in the United States.

Do Companies Really Need Registration?

The advantage of a third-party audit is that it assures an objective assessment and verification that the standards have been adopted, assuming it is performed by an accredited auditor. However, registration (accreditation) can be very costly, and just because a company has the ISO 14001 certification, there are no guarantees of good environmental performance or in reaping the many benefits of having an EMS (such as compliance, P2 cost savings, better management, and control over operations).

For this reason, many companies opt for self-declaration. One example of a self-declaration program is that of the chemical industry, the Chemical Industry's Responsible Care®-14001 (Rc®-14001) program. In short, EMS benefits can be obtained without certification, but self-declaration is not enough unless the EMS is truly established within the organization.

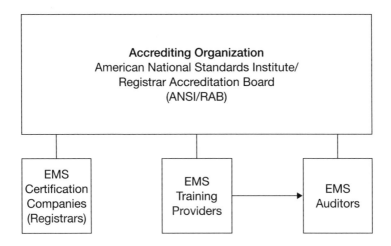

For registration (certification) to ISO 14001:
1. Enterprise requests (and pays for) audit by RAB-accredited registrar (registrar may be accredited by a counterpart organization in another country).
2. Registrar hires, fields, and supervises a team of RAB-certified ISO 14001 auditors, including a RAB-certified lead auditor (auditors may be certified by a counterpart organization in another country).
3. RAB-certified auditors must have successfully completed training by a RAB-accredited training provider and meet other requirements, including annual recertification requirements (auditing hours, professional development, etc.) (auditor training may be provided by a counterpart organization accredited in another country).

Figure 1-3 ISO 14001 accreditation and certification infrastructure.

Justifying the EMS

The case studies presented in this chapter should convince most readers that simply meeting environmental obligations is not adequate in achieving and maintaining good environmental performance. Even when a company does not take its environmental performance seriously, it most certainly should be sensitive to the potential liabilities from not acting responsibly and being able to demonstrate a level of performance that goes beyond compliance.

If nothing else, an EMS focuses on continually mitigating the negative environmental aspects of a business. Through the continual improvement cycle, a company dedicates resources to address the environmental aspects of the business all the time. It is continually planning, doing, checking, and revising its protocols, practices, and operations in ways that focus on systematically eliminating emissions and wastes. On a conceptual basis alone, this is an infinitely better approach to meeting present, changing, and future regulatory obligations than having an environmental department that simply reacts to problems and plays catchup in adapting its practices and control technologies when regulations change.

There are also direct benefits that affect on the financial wellness of a business. Among those recognized are

- Assuring customers of commitment to demonstrable environmental management.
- Maintaining good public and community relations.
- Satisfying investor criteria and improving access to capital.
- Obtaining insurance at reasonable cost.
- Enhancing image and market share.
- Meeting vendor certification criteria.
- Improving cost control.
- Reducing incidents that result in liability.
- Demonstrating reasonable care.
- Conserving input materials and energy.
- Facilitating the attainment of permits and authorizations.
- Fostering development and sharing environmental solutions.
- Improving industry-government relations.

2
Getting an EMS Started

Introduction

This chapter focuses on the practical aspects of getting an EMS off the ground. As noted in the Preface, it is not the intent of this volume to provide an extensive review of the almost overwhelming volume of literature on environmental management systems, nor is it my intent to rehash earlier publications on the subject. Rather, this book may be viewed as a supplement to both the already rich information and tools available for getting an EMS off and running. To some readers, the volume will serve as a primer, whereas others will find this to be either a review that reinforces their company's program or a starting basis for establishing their EMS.

This chapter ties together many of the important elements needed at the starting gate for establishing an EMS. While ISO 14001 is the benchmark EMS discussed throughout the volume, this chapter provides concepts and tools that are relevant to any EMS. The reader will find a toolkit at the end of the chapter that will facilitate the initial step in the process of an EMS implementation.

Practical Aspects of an EMS

According to the International Organization for Standardization an EMS is "that part of the overall management system which includes organizational structure, planning activities, responsibilities, practices, procedures, processes and resources for developing, implementing, achieving, reviewing and maintaining the environmental policy." As noted in Chapter 1, the elements of an EMS are

- *Policy.* Top management needs to clearly define the organization's environmental policy. This written policy statement can apply to facilities, sites, and programs, appropriate to the scale of operation and impact of activities governed by the policy. The format and length can vary, but the policy must commit to relevant environmental legislation and regulations. It must provide a framework for setting and reviewing goals. The policy needs to be communicated to all employees, the public, and in fact along the entire supply chain that interacts with a business. Referring back to Figure 1–2 in Chapter 1, the very first step in the implementation of ISO 14001 involves Clause 4.2, which recommends that a company make a commitment to *pollution prevention* as its top priority. There are no "boilerplate" policy statements and indeed these statements can range from very general commitments to complying with environmental regulations and protecting the environment by reducing waste and pollution to statements that reflect very specific targets and goals. The environmental policy statement, in fact, can reflect not just an overall policy but rather be viewed as vision statement with specific targets and goals to achieve. Consider some of the following statements published by well-known companies that have been downloaded from the WWW:

    ENVIRONMENTAL HEALTH AND SAFETY (EHS) EFFECTS ANALYSIS by every facility and business operation, to identify their most significant EHS impacts and opportunities for improvement (by year-end 2003 for manufacturing and research and development facilities and year-end 2004 for major commercial operations).

SAFETY PERFORMANCE TARGETS to drive down reportable incidents and days lost from work.

- Fifty percent reduction in recordable and lost workday case rates based on 2000 base year.
- Maintain performance in top 25 percent of similar companies.

ENVIRONMENTAL PERFORMANCE TARGETS AND GOALS (normalized by sales) to reduce water use, greenhouse gas emissions, energy use, wastewater, air emissions, hazardous waste, nonhazardous waste, and off-site releases of chemicals of high concern.

1. Reduce energy use by 10 percent from 2001 baseline year.
2. Reduce water use by 10 percent from 2001 baseline year.
3. In countries where water resources are severely stressed, develop a plan by 2003 with 2010 stretch goals for additional water-use reduction.
4. Reduce total greenhouse gas emissions (e.g., carbon dioxide and methane) by 10 percent from 2001 baseline year.
5. Establish company-wide reduction goals for nonhazardous waste by 2003, based on results of EHS effects analysis at each facility.
6. Reduce off-site hazardous waste disposal by 50 percent from 2001 baseline year.
7. Establish company-wide reduction goals for air emissions of acid gases (sulfur oxides, nitrogen oxides, and hydrogen chloride) by 2003, based on results of EHS effects analysis at each facility.
8. Reduce off-site release to air of priority reduction chemicals by 50 percent from 2002 baseline year.
9. Establish company-wide reduction goals for the following wastewater releases by 2003, based on results of EHS effects analysis at each facility: total chemical oxygen demand, suspended solids, nitrates, and heavy metals.
10. Reduce off-site release to water of priority reduction chemicals by 50 percent from 2002 baseline year.

These examples reflect very specific goals and targets and are embedded within the general policy statements that commit to acting in a responsible manner for public safety, workforce safety, conserving resources, protecting the environment, and meeting compliance. They in turn are a reflection of core values relating back to good business practices that have to do with being profitable, competitive, sustainable, and acting on a responsible manner. Table 2–1 is a checklist that can aid in formulating the environmental policy statement for your company.

- *Planning*. Planning begins with identifying the environmental aspects of activities the organization controls (i.e., the components of those activities that are likely to interact with the environment) and understanding how those aspects have an impact on the environment. In a practical sense, once the policy and commitment by top management are established and the mes-

Table 2-1 Environmental Policy Checklist

To what extent do the following statements apply to your policy?					
Not at all = 1 point; absolutely = 5 points	1	2	3	4	5
1. The environmental policy of your company is written	☐	☐	☐	☐	☐
2. It is defined by management	☐	☐	☐	☐	☐
3. Employees participated in defining the policy	☐	☐	☐	☐	☐
4. It is reviewed at regular intervals	☐	☐	☐	☐	☐
5. When changes occur in the company, it is adapted	☐	☐	☐	☐	☐
6. Employees are informed about the policy	☐	☐	☐	☐	☐
7. It covers compliance with environmental legislation	☐	☐	☐	☐	☐
8. It covers commitment to continuous improvement of environmental performance	☐	☐	☐	☐	☐
9. The following site-related areas are taken into account:					
a. Energy	☐	☐	☐	☐	☐
b. Raw materials and water	☐	☐	☐	☐	☐
c. Waste	☐	☐	☐	☐	☐
d. Noise	☐	☐	☐	☐	☐
e. Production processes	☐	☐	☐	☐	☐
f. Product planning	☐	☐	☐	☐	☐
g. Environmental performance of contractors and suppliers	☐	☐	☐	☐	☐

sage becomes clear throughout an organization, the company needs to put its arms around the environmental aspects to move forward with implementing an EMS. While there are formal definitions floating around, I devised one that helps to integrate cleaner production/pollution prevention (CP/P2) more readily into the EMS. An *environmental aspect*, or EA, can be defined as the interactions other than those related to or created by the final product between any unit process, operation, practice, piece of equipment, or feature of a production (manufacturing) operation and the surroundings. The role of the planning stage of an EMS is to identify the negative environmental aspects and devise ways to eliminate them or minimize their negative impacts. In short, the planning stage of an

Table 2-1 *continued*

Not at all = 1 point; absolutely = 5 points	1	2	3	4	5
h. Prevention of accidents	☐	☐	☐	☐	☐
i. Dealing with accidents	☐	☐	☐	☐	☐
j. Environmental protection and personnel	☐	☐	☐	☐	☐
k. Environment and public relations	☐	☐	☐	☐	☐
10. The policy is understandable for employees	☐	☐	☐	☐	☐
11. It is understandable for those outside of the company	☐	☐	☐	☐	☐
12. It is goal oriented	☐	☐	☐	☐	☐
13. It has well-defined targets for improvement	☐	☐	☐	☐	☐
14. It is credible	☐	☐	☐	☐	☐
15. It serves as a guide for employees	☐	☐	☐	☐	☐
16. It helps employees to identify with the company	☐	☐	☐	☐	☐
17. It is followed up with concrete measures and goals	☐	☐	☐	☐	☐
18. It fits with the ability of management and employees	☐	☐	☐	☐	☐
19. It is consistent with the overall goals of the company	☐	☐	☐	☐	☐
20. It is supported and put into practice by management	☐	☐	☐	☐	☐
21. It incorporates the needs of day-to-day operations and opens up possibilities as opposed to being a constraint	☐	☐	☐	☐	☐

Total ___
Maximum no. of points possible   (100%) = 155 points
Score obtained   ___ (%) = ___ points

EMS directs our attention to determining as many, if not all, of the negative EAs possible. Once EAs and their impacts have been identified, criteria can be applied to rank or prioritize them (see Figure 2–1). This establishes a rational decision-making basis for systematically reducing the negative impacts on a schedule consistent with the business goals and longer-term strategies that address such crucial issues as growth and sustainability of the company as a whole. By defining and establishing a priority as to which EAs should be addressed first, based on well-defined criteria, *top management* can then establish objectives, goals, and targets requirements. This allows the company to define *environmental action plans* for each of the EAs and place them on a schedule for resources and implementation. The process or activity by which EAs are defined is known as an *initial environmental review*, or IER. Examples of how the EAs are ranked are provided in the book *Green Profits*.[1]

- *Implementation and operation.* As the term implies, this step in the implementation of an EMS is putting the environmental action plans into effect. This step, however, involves not just implementing corrective actions or changes that reduce and eliminate EAs but also embraces programs for training and awareness, establishing avenues for communication inside and outside the organization, maintaining documentation, and planning for operational control and emergency response. Just some of the key issues that need to be addressed are pollution prevention, continuous improvement, and compliance for reducing identified impacts and the development of managerial programs for achieving them, including a mechanism for identifying applicable legal and other issues that encompass defining roles and responsibilities.

---

[1] N. P. Cheremisinoff and A. Bendavid-Val, *Green Profits: A Manager's Handbook to ISO 14001 and Pollution Prevention* (Oxford: Butterworth-Heinemann Publishers, 2001).

Getting an EMS Started

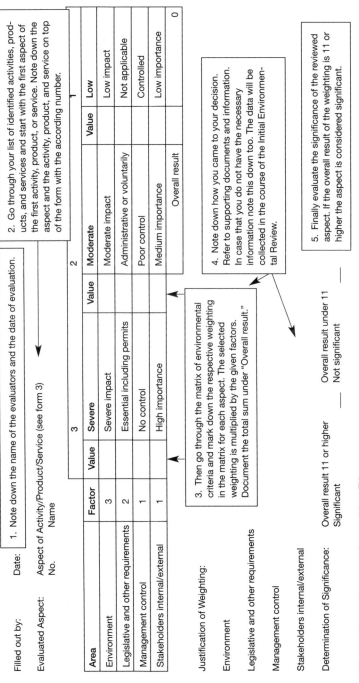

Figure 2–1 Protocol for ranking EAs.

- *Checking and corrective action.* An organization must measure its performance against its own targets and objectives, its operational controls, and its compliance with relevant laws and regulations. Specifically, an EMS must define how nonconformance with the ISO standard will be handled and how corrective measures will be taken. In addition, checking and corrective action is the basis by which continual improvement of environmental performance is achieved. From this standpoint, it is also the basis that renews top management's commitment to the EMS and further refines the policy and vision of a company. This, unto itself, is not a simple task, because the larger and more complex an organization is, the more EAs there are to manage. Furthermore, actions have cost impacts. And, in a practical sense, that is what an EMS reduces to—controlling the costs, whether present day or based on future liabilities and impacts of environmental management. Senior management must have a clear understanding that P2 and corrective actions from environmental action plans do not simply reduce emissions and noncompliance but also capture financial savings, whether it be from reductions achieved in reduced pollution fees or from improved efficiencies, operational performance, energy demand reductions, raw material use savings, improvements to product quality and labor efficiency, and others. Figure 2–2 illustrates what many companies often track. Overwhelmingly, the category of environmental benefits are tracked and reported by companies that have embraced an EMS. But, to business leaders and decision makers, environmental benefits are most readily understood in terms of dollars. In fact, for investors, lending institutions, insurance carriers, and partners, the dollar savings associated with good environmental performance are related best to using performance tracking as a means of stewardship when it comes to environmental management. In devising the proper approach to checking and corrective action, the proper *metrics* need to be defined, and an appropriate *environmental management information system,* or

EMIS, must be selected and applied. These terms are discussed further on in this chapter.

- *Management review.* To build in continuous improvement, top management must periodically review the system and address needed changes. This starts the process of continual improvement all over again.

Thus plan, do, check, and revise are the basic elements of any EMS. Environmental management systems can add significant value for any company, large or small, in a variety of ways. For example, they can

- Provide a cost-effective corporate environmental framework.
- Reduce risks of regulatory noncompliance by systematically tracking applicable requirements.
- Provide a basis for discussing flexibility in regulatory interpretations.
- Help maintain stakeholder confidence by demonstrating reliable environmental protection.

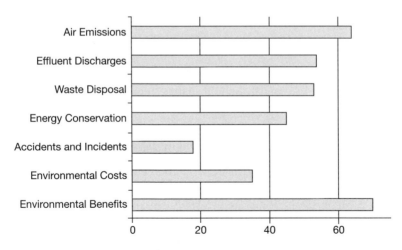

Figure 2-2 Metrics most often tracked and reported by companies.

- Provide a basis for self-correction and mitigating fines and penalties if noncompliance does occur.
- Integrate the principles and cost savings of pollution-prevention activities.
- Help programs to focus on the most important environmental aspects.
- Provide an integrating framework for EHS (environmental health and safety) activities.
- Facilitate deployment of new technologies at sites.
- Provide a basis for "benchmarking" against competitors, within industry sectors and against sector operations.
- Provide vision and stability for staff members and decision makers in turbulent times.
- Demonstrate responsibility, accountability, and continuous improvement in managing a company's sites, programs, and assets.
- Bolster investor and public confidence, which can further provide a competitive edge.
- Demonstrate environmental leadership across the private sector.

Implementing an EMS does not alter the basic obligation to comply with applicable requirements or the requirements themselves. Moreover, environmental management systems, by themselves, do not guarantee the effectiveness of the management system itself. Although compliance is already a requirement, an EMS provides ways to make achieving that end more sustainable. For example, an EMS requires a mechanism for systematically identifying applicable legal requirements.

What an EMS does is enable a company to focus on making its operations more efficient by being proactive rather than reactive to the EAs of the business. An EMS is designed to continuously improve the management of environmental activities at a facility, to avoid gaps and lapses in compliance when requirements come into play, and to curtail future negative impacts from changing and ever-tightening regulations that command greater levels of responsible care. An EMS provides the overall management framework within which the activities will be conducted. Thus, an EMS can be used to deploy or amplify the effectiveness of other EHS initiatives.

As discussed in Chapter 4, an EMS can help managers and an entire organization work smarter, faster, cheaper, and accomplish more with less. Improved management leads to more efficient and effective performance and allows greater control over environmental support costs.

A planned comprehensive periodic audit of the EMS is important to ensure that it is effective in operation, meets specified goals, and continues to perform in accordance with relevant regulations and standards. The audits are designed to provide additional information in order to exercise effective management of the system, providing information on practices that differ with the current procedures or offer an opportunity for improvement.

In addition to the audit, the requirement for management review of the system ensures that it is suitable (for the organization and the objectives) and effective in operation. The management review is the ideal forum to make decisions on how to improve for the future.

Examples of EMSs

The International Standards Organization (ISO) developed a series of standards and guidelines in the field of environment, which collectively are known as the ISO 14000 series. It is not correct to refer to the ISO 14000 standard. ISO 14001 is the only certifiable standard, the remainder being supportive guidelines. The various standards and guidelines in the series are listed in Table 2–2.

ISO 14001 is one type of EMS. There are many types of EMSs. Some have been developed by industry associations, such as the chemical industry's *Responsible Care*, the hospitality industry's *Green Globe*, and the forest industry's *SmartWood*. Some nongovernmental

Table 2-2 Standards and Guidelines within the ISO 14000 Standards

Series	Subject	Standards
ISO 14000	Environmental management systems	ISO 14001 ISO 14004
ISO 14010	Guidelines for environmental auditing	ISO 14010 ISO 14011 ISO 14012
ISO 14020	Environmental labels and declarations	ISO 14020 ISO 14021 ISO 14022
ISO 14030	Evaluating environmental performance	ISO 14031
ISO 14040	Life-cycle analysis	ISO 14040 ISO 14041 ISO 14042 ISO 14043
ISO 14050	Understanding terms and definitions	ISO 14050

organizations (NGOs) developed environmental management systems, such as the International Chamber of Commerce's *Business Charter for Sustainable Development*. Government agencies also developed EMSs for their own purposes; examples are the *Facility Environmental Management and Monitoring System*, of the Tobyhanna Army Depot (www.femms.com) and the North American Commission for Environmental Cooperation (NAFTA associated) *EMS Guidelines* (www.cec.org). And, of course, many individual enterprises designed their own EMSs: ExxonMobil's *OIMS* (operational integrity management system for EHS) is one example, and Baxter International's well-known EMS based on its own adaptation of the ISO 14001 model is another. All these, in some manner, follow the logic of the basic plan-do-check-revise/act model. The following are general EMSs that exist within the European Community.

Eco-Management and Audit Scheme

Eco-Management and Audit Scheme (EMAS) was introduced by a European Union council regulation (No. 1836/93), requiring implementation in all European Union member states. The EMAS regulation was first introduced in July 1993 as an environmental policy tool devised by the European Commission, in a step toward the community's goal of sustainable development. The EMAS scheme was open for voluntary participation by organizations from April 1995 and its scope restricted participation to sites operating industrial activities.

In 1996, the international environmental management system standard, EN ISO 14001, was published and recognized as a step toward achieving EMAS. It was also recognized that all sectors have significant impacts on the environment and the environment benefits from good environmental management in other sectors as well.

Article 14 of the regulation allowed member states to extend the scheme to other economic sectors, and several member states took the opportunity to pilot EMAS successfully in those other sectors. Article 20 of Regulation 1836/93 stated that the EMAS scheme had to be reviewed no more than five years after its entry into force.

In 1997, the consultation process began, which took these changes into consideration. The commission adopted a proposal that went through the codecision procedure, involving other European institutions (Economic and Social Committee, Council of the Regions) and giving the European Parliament similar rights to the council in the decision process.

In 2001, the new Regulation (EC) No. 761/2001 was adopted (O.J. L114, 24.4.2001, p. 1). The main elements are:

- Extension of the scope of EMAS to all sectors of economic activity, including local authorities.
- Integration of ISO 14001 as the environmental management system required by EMAS, so that progressing from ISO 14001 to EMAS will be smoother and not entail duplication.
- Adoption of a visible and recognizable EMAS logo to allow registered organizations to publicize their participation in EMAS more effectively.
- Involvement of employees in the implementation of EMAS.

- Strengthening of the role of the environmental statement to improve the transparency of communication of environmental performance between registered organizations and their stakeholders and the public.
- More thorough consideration of indirect effects, including capital investments, administrative and planning decisions, procurement procedures, choice and composition of services (e.g., catering).

The regulation consists of 18 articles and 8 annexes. Unlike other management systems standards, the annexes form part of the regulation and are not merely informative. This means that the requirements of the annexes must be met.

The overall objective of the European Community, as defined by the Maastricht Treaty, is to "promote a harmonious and balanced development of economic activities, sustainable and non-inflationary growth respecting the environment . . . the raising of standards of living and quality of life."

The scheme does not replace existing community or national environmental legislation or technical standards nor does it remove a company's responsibility to fulfill all its legal obligations under such legislation or standards.

The goal of sustainable development calls for the use of a wider range of tools for environmental policy. The Sixth Community Environmental Action Program, Environment 2010: Our Future, Our Choice, recognizes this and aims to be a program that "completes and reinforces our body of environmental legislation where there are gaps and takes forward the implementation of our directives . . . does more in terms of mobilizing stakeholders for the environment and 'greening' the market."

Originally, the EMAS scheme was site based and open only to companies operating industrial activities. However, the EMAS scope has widened, and it can now be applied to any organization that has an impact on the environment. Participation in the scheme is open to organizations operating in all economic sectors.

The EMAS identifies the different types of economic activity areas by using the "Statistical Classification of Economic Activity (NACE) Codes." NACE stands for Nomenclature générale des activités économiques dans les Communautés Européennes, which is the standard for classification of economic activities in the European Union. Details of these can be found in Council Regulation No. 3037/90 (as amended by Regulation No. 761/93 and 29/2002), NACE Rev 1. Under the previous regulation, the scheme was open to the industrial sector covering only NACE codes from 10 to 40. Now that EMAS II allows participation of organizations from all economic sectors, the whole range of NACE codes is applicable. This list enumerates NACE codes with a maximum of two decimals after the comma; however, some NACE lists go much deeper into detail with up to five decimals after the comma.

For EMAS, the NACE codes are used to classify the registered organizations and the accredited verifiers. The agreement was made that the "competent body" that registers the site and forwards the information to the EMAS Helpdesk should collect NACE information with a minimum of one decimal after the point. It varies from one competent body to the other if they go into more detail or not.

Registration in the scheme requires an organization or company to adopt an environmental policy containing commitments both to comply with all relevant environmental legislation and to achieve continuous improvement in environmental performance.

At the company or organization, an initial environmental review is undertaken. In the light of this review and the policy, an environmental program and environmental management system are established.

The EMAS is generally described as being more prescriptive than ISO 14001 and goes beyond the ISO standard. To a degree, this is arguable and often places the two systems in competition. I argue neither one way nor the other but note that, from the standpoint of a company acting responsibly, there is little difference as to which standard is chosen. Notable differences between the two include these areas:

- *Preliminary review.* Europeans argue that EMAS requires a verified initial environmental review, whereas ISO does not. However, ISO 14001 simply cannot be implemented without an IER. Since a company undertakes the investment of time and resources to implement an EMS, performing an IER using internal experts who best know and understand their operations is more verifiable than relying on external auditors.

- *Public availability.* EMAS requires that the policy, program, environmental management system, and details of the organization's performance are made publicly available as part of the environmental statement. ISO requires only that the policy be publicly available. But, in a practical sense, when companies demonstrate good environmental performance, they widely publicize their performance. Few companies ignore environmental metrics and the power and rewards derived from reporting improved performance.

- *Audits.* ISO 14001 requires audits, although the frequency is not specified nor is the audit methodology set out in as much detail as in the EMAS.

- *Contractors and suppliers.* The EMAS is slightly more explicit in its control over contractors and suppliers, requiring that procurement issues are addressed and the company endeavors to ensure that contractors and suppliers comply with the environmental policy. ISO 14001 requires that relevant procedures are communicated to contractors and suppliers. Some companies actually require their preferred suppliers to adopt ISO 14001 standards. In effect, there should be no difference.

- *Commitments and requirements.* ISO 14001 does not stipulate the extent to which performance must be improved. The EMAS specifies that organizations must attempt to "reduce environmental impacts to levels not exceeding these corresponding to economically viable application of best available technology."

Perhaps, the largest distinction between EMAS and ISO 14001 is certification. To comply with the requirements of EMAS, the ISO 14001 certificate must be issued under one of the accreditation procedures recognized by the European Commission. While certification (or registration) is the same in the United States, it is not mandatory. ISO 14001 is voluntary, and as such, many companies elect for self-declaration to minimize costs. While critics argue this invalidates the EMS, the fact is that, if a company truly implements an EMS, the formal registration process using independent auditors and consultants will not provide a whole lot more except cost. And, for those companies that have recently undergone ISO 9001 certification and see the value of a third-party implemented program, costs can be kept down by utilizing the same consultants for an ISO 14001 exercise.

I have consulted at facilities outside the United States where ISO 14001 registration already existed. Despite this, the environmental management practices at these facilities were nowhere close to the self-declared programs found in the United States. One important reason for this is that nonfunctioning but registered EMSs were observed in countries with lax environmental enforcement. Since a major driver for improved environmental performance in North America are strict and aggressively enforced environmental laws, the EMS is taken seriously, regardless of whether it is registered or not. Another very important reason why registered EMSs that are nonfunctional can be found outside the United States and the European Community is the role of privatization. In countries with transitioning economies and third-world nations, major enterprises are often government owned or heavily subsidized. Under these circumstances, the incentives for cost savings are greatly diminished; hence, there are fewer incentives for a company to invest in developing and maintaining a working EMS. Enterprises that have obtained ISO 14001 registration often do so because of a perception that it can open up foreign markets.

EMAS implementation steps for an ISO 14001 certified facility are outlined in Figure 2–3.

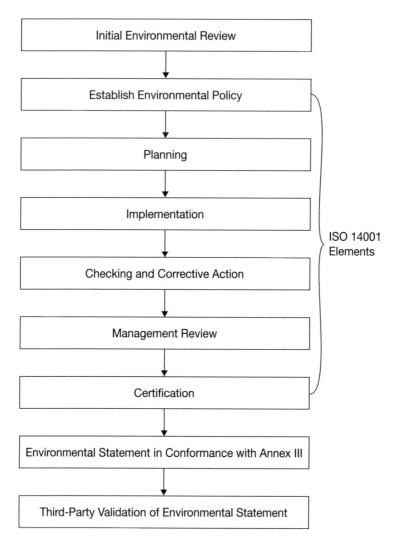

Figure 2-3 EMAS implementation route to ISO 14001 certification.

British Standard 7750

The United Kingdom has had its own EMS Standard, BS7750, since 1992 and increasing numbers of companies, within and outside the United Kingdom, have achieved registration to it. The requirements of BS7750 differ slightly from ISO 14001.

BS7750 is used to describe a company's environmental management system, evaluate its performance, and define policy, practices, objectives, and targets and provides a catalyst for continuous improvement. The concept is similar to the use of ISO 9000 for high-quality systems, with the methods to be used open to definition by the company. The standard provides the framework for development and assessment of the resultant environmental management system.

BS7750 was developed as a response to concern about environmental risks and damage (both real and potential). Compliance to the standard is voluntary for companies and complements required compliance to statutory legislation.

BS7750 is designed to be compatible with the European Community's EMAS and ISO 14001.

BS7750 requires an environmental policy to be in existence within the company, fully supported by top management, and outlining the policies of the company, not only to the staff but to the public, just like ISO 14001. The policy needs to clarify compliance with environmental legislation that may effect the organization and stress a commitment to continuous improvement. Emphasis has been placed on policy, as this provides the direction for the remainder of the management system.

The preparatory review and definition of the organization's environmental effects is not part of a BS7750 assessment; however, examination of this data provides an external audit with information on the methods adopted by the company. The preparatory review itself is comprehensive in consideration of input processes and output at the site. The review (an IER) is designed to identify all relevant environmental aspects that may arise from existence on the site. These may relate to current operations; they may relate to future, perhaps even unplanned future, activities; and they will certainly relate to the activities performed on-site in the past (i.e., contamination of land, groundwater, etc.).

The IER also includes a wide-ranging consideration of the legislation that may affect the site, whether it is currently being complied with and perhaps even whether copies of the legislation are available.

The company is required to declare its primary environmental objectives, those that can have most environmental impact. To gain the most benefit, these become the primary areas of consideration within the improvement process and the company's environmental program. The program embodies a plan to achieve specific goals or targets along the route to a specific goal and describes the means to reach those objectives such that they are real and achievable. The environmental management system thus provides further detail on the environmental program.

All EMSs establish procedures, work instructions, and controls to ensure that implementation of the policy and achievement of the targets can become a reality. Communication is a vital factor, enabling people in the organization to be aware of their responsibilities, aware of the objectives of the scheme, and able to contribute to its success.

Defining Pollution Prevention and Related Terminology

The following is a passage from the Pollution Prevention Act of 1990 that the U.S. Environmental Protection Agency (EPA) developed as a formal definition of pollution prevention and a strategy for making pollution prevention a central guiding mission. Under Section 6602(b) of the Pollution Prevention Act, Congress established a national policy that

- pollution should be *prevented* or *reduced* at the source whenever feasible;
- pollution that cannot be prevented should be *recycled* in an environmentally safe manner whenever feasible;
- pollution that cannot be prevented or recycled should be *treated* in an environmentally safe manner whenever feasible; and
- disposal or other release into the environment should be employed only as a last resort and should be conducted in an environmentally safe manner.[2]

---

[2]Henry F. Habicht II, *Memorandum: EPA Definition of Pollution Prevention* (Washington, DC: U.S. Environmental Protection Agency, May 28, 1992).

This hierarchy of preferred options for dealing with environmental pollution officially places prevention first.

According to the EPA's official definition, *pollution prevention* means "source reduction" as defined in the Pollution Prevention Act. However, the term *pollution prevention*, or P2, also includes "other practices that reduce or eliminate the creation of pollutants through:

- increased efficiency in the use of raw materials, energy, water, or other resources, or
- protection of natural resources by conservation."

The term *source reduction* is defined under the act as any practice that

- Reduces the amount of any hazardous substance, pollutant, or contaminant entering any waste stream or otherwise released into the environment (including fugitive emissions) prior to recycling, treatment, or disposal; and
- Reduces the hazards to the public health and the environment associated with the release of such substances, pollutants, or contaminants.

Source reduction includes "equipment or technology modifications, process or procedure modifications, reformulation or redesign of products, substitution of raw materials, and improvements in housekeeping, maintenance, training, or inventory control."

P2 may be thought of as roughly synonymous with source reduction; that is, reducing the generation of wastes or contaminants at the source and thereby reducing releases to the environment that could pose hazards to the environment and public health. Like source reduction, P2 as defined by the Pollution Prevention Act does not include out-of-process recycling, waste treatment, or combustion of wastes for energy recovery.

Recycling has been excluded from the official definition of P2, because in essence, it is not a form of prevention. However, recycling can confer substantial environmental improvements and contribute to conserving valuable resources. Many experts in the industry argue that recycling should be on par with P2, since it represents progress toward reducing environmental pollution and achieving greater efficiency in resource use. But, from a purist standpoint, recycling should be excluded from the definition of pollution prevention because even wastes that are effectively recycled have not been *prevented*. Recycling, however, does rank as the second highest option in the EPA's pollution prevention/waste management hierarchy, which attests to the desirability as a goal in cases where wastes cannot be feasibly prevented. In addition, in some cases, in-process recycling can be considered a form of P2, when materials are directly reincorporated back into the same process.

Source reduction and P2 are essentially synonymous with each other, and in fact, it is an industry-derived term that likely significantly predates P2.

P2 itself is a term that can have a variety of meanings, depending upon who is using it. Although the EPA's definition is perhaps the most widely known, others have defined *pollution prevention* to include recycling and reclamation activities (activities that Congress and the EPA specifically exclude). For example, a draft standard being prepared by the American Society for Testing and Materials (ASTM) on the development and implementation of pollution prevention programs defines *pollution prevention* as "the act of reducing or eliminating the use, release or generation of a pollutant or potential pollutant through source reduction, recycling, reuse, reclamation or modification of existing practices."[3]

Another very commonly used term is *waste minimization* or *waste min*. Waste min was among the first initiatives in the area of pollution prevention. It focused almost exclusively on solid wastes regulated under the Resource Conservation and Recovery Act (RCRA) and, in particular, hazardous wastes.[4] Waste minimization is much narrower than pollution prevention, as the latter focuses on

---

[3] ASTM E50.03 Subcommittee on Pollution Prevention, Reuse, Recycling and Environmental Efficiency, *Standard E50.03.1: Guide for Development and Implementation of a Pollution Prevention Program*, working document, January 24, 1994. The standard is available from the ASTM Customer Service Department by calling (215)299-5585.

[4] U.S. EPA, *Pollution Prevention 1991: Progress on Reducing Industrial Pollutants*, EPA 21p-3003 (Washington, DC: Office of Pollution Prevention, U.S. EPA, October 1991), pp. 6–7.

reducing the entire spectrum of pollution and waste, including air emissions, releases to surface and groundwaters, and inefficient energy and materials use, in addition to waste (in the traditional sense), which is sent off for land disposal, treatment, or off-site recycling. Also waste minimization practices have been controversial, since they include treatment methods to reduce the volume or toxicity of wastes as opposed to focusing solely on eliminating waste being generated at the source. The RCRA reporting requirements exclude treatment and energy recovery from the definition of waste minimization activities. However, unlike the EPA's definition of *pollution prevention*, waste minimization does include recycling in addition to source reduction activities.[5]

Waste reduction is another term that industry likes to throw around, but this term falls somewhere between waste minimization and P2. Waste reduction has a broader focus than waste minimization, with its emphasis on RCRA hazardous wastes, but implies a narrower perspective than P2, with its holistic approach to preventing all types of pollution released to all environmental media from products as well as industrial processes.

Toxics use reduction is the elimination or avoidance of using toxic substances in products or processes so as to reduce the risks to the health of workers, consumers, and the general public and minimize adverse effects on ecosystems and the environment. Toxics use reduction falls under source reduction. *Toxic chemical use substitution* refers to the substitution of toxic chemicals with less harmful substances in products or processes. It can also include efforts to reduce or eliminate the use of specific chemicals or categories of toxic substances through the development of appropriate substitutes or alternative technologies. Source reduction and toxic chemical use substitution together make up industrial pollution prevention.[6]

---

[5]Henry Freeman et al., "Industrial Pollution Prevention: A Critical Review," *Journal of Air and Waste Management* 42, no. 5 (May 1992): 619–620.

[6]U.S. EPA, *Pollution Prevention 1991: Progress on Reducing Industrial Pollutants.* EPA 21p-3003 (Washington: Office of Pollution Prevention, U.S. EPA, October, 1991), pp. 6–7.

Eco-efficiency is a term that was coined by the World Business Council for Sustainable Development (WBCSD) in 1992 and defined as the delivery of competitively priced goods and services that satisfy human needs and bring quality of life, while progressively reducing ecological impacts and resource intensity throughout the life cycle, to a level at least in line with the earth's estimated carrying capacity. Concepts of eco-efficiency and P2 are in many ways almost synonymous. The slight difference between them is that eco-efficiency starts from issues of economic efficiency that have positive environmental benefits, while P2 starts from issues of environmental efficiency that have positive economic benefits.

Green productivity is a term used by the Asian Productivity Organization (APO) to address the challenge of achieving sustainable production. The APO started its Green Productivity program in 1994. Like P2, green productivity is a strategy for enhancing productivity and environmental performance for overall socioeconomic development.

Industrial ecology and *industrial metabolism* are concepts for new patterns of industrial production and closely related to the P2 concept. Industrial ecology and industrial metabolism are studies of industrial systems and economic activities and their links to fundamental natural systems. They aim to imitate the material recycling aspect of an ecosystem, a material flow management is the crucial aspect of these approaches. There are six elements of industrial ecology/industrial metabolism:

1. The creation of industrial ecosystems—maximizing use of recycled materials in production, optimization of the use of materials and embedded energy, minimizing waste generation, and reevaluating wastes as raw material for other processes.

2. Balancing industrial input and output to natural ecosystem capacity—understanding the ability of the larger natural system to deal with toxins and other industrial wastes in typical and catastrophic situations.

3. Dematerialization of industrial output—reducing materials and energy intensity in industrial production.

4. Improving the metabolic pathways of industrial processes and materials use—reducing or simplifying industrial processes to emulate natural, highly efficient ones.
5. Systemic patterns of energy use—promoting the development of an energy supply system that functions as a part of the industrial ecosystem and is free of the negative environmental impacts associated with current patterns of energy use.
6. Policy alignment with a long-term perspective of industrial system evolution—nations working together to integrate economic and environmental policies.

The term *cleaner production* is essentially interchangeable with P2, although I have run across opinions from other experts that make slight distinctions. But, from a practical sense, the only real distinction between the two terms tends to be largely one of geography. The term *pollution prevention* tends to be used in North America, while *cleaner production*, or CP, is used in other parts of the world. Both CP and P2 focus on a strategy of continuously reducing pollution and environmental impacts through source reduction. Both focus on eliminating waste within the process rather than at the end of the pipe. Some definitions quoted from the WWW are

- "Cleaner production (CP) is the continuous application of an integrated preventive environmental strategy applied to processes, products and services to increase efficiency and reduce risk. This strategy typically involves the modification of production processes, using a life-cycle approach, and results in meeting customer needs with more environmentally compatible products and services. Cleaner production also brings tangible economic savings and financial benefits. Parallel concepts to cleaner production are pollution prevention, waste minimization, eco-efficiency and green productivity."[7]
- "Means avoiding the use or production of environmentally persistent substances, reducing the amount or toxicity of

---

[7] www.lineadecreditoambiental.org/html/glossary.html.

liquid wastes produced that require disposal and reducing the amount or toxicity of solid wastes."[8]

- "Any manufacturing or production process which assist with the reduction of waste, the reduction emissions and energy reduction."[9]
- "Improvements to a production process so the process uses less energy, water or other input, or generates less waste or less environmentally harmful waste (adapted from the Waste Environmental Protection Policy)."[10]

In all discussions hereafter, we use the terms CP/P2, meaning cleaner production and pollution prevention. There are simply those who feel one term is more appropriate than the other, but they both have the same objective. That objective is to "prevent" as opposed to react and control.

Pollution Prevention within the Context of an EMS

Many North American companies justify the investment in an EMS to minimize liabilities associated with noncompliance. As pointed out in my prior publications, a variety of benefits are derived from an EMS, for which a discussion is provided in Chapter 4. For now, let us simply say that CP/P2 is a main driver in achieving the many benefits, including improved compliance. From a more general standpoint, CP/P2 does not just focus on regulated wastes or pollution but all forms of wastes. In a perfect world, 100% of the raw materials, energy, and labor input needed to manufacture a useful product would be transformed into that product. Limitations in hardware, technology, engineering, and human error result in only partial conversion of raw materials and energy into useful products demanded by society. The remainders are by-products and lost

---

[8]www.waitakere.govt.nz/glossary.asp.

[9]www.wastenot.ie/glossary.html.

[10]www.epa.qld.gov.au/environmental_management/sustainability/industry/sustainability_roadmap/glossary/.

energy. Some of the by-products are hazardous, while others are not, yet they still are among the waste streams that must be dealt with. All unusable by-products, including lost energy, are forms of pollution. In fact, the terms *pollution* and *waste*, as used in this volume, are interchangeable. The aim of CP/P2 is to find economical ways to prevent the formation of the by-products, lost energy, inefficiencies, reduce human error, and minimize the use of raw materials per unit of production. Clearly, the more one is able to reduce "pollution," the more efficient a manufacturing operation becomes, and hence the more profitable the company becomes.

By applying CP/P2 practices, technologies, and attitudinal practices among management and employees in managing pollution, the dependence on "end-of-pipe" solutions can be reduced or even eliminated altogether. CP/P2 has been applied to raw material extraction, manufacturing, agriculture, fisheries, transportation, tourism, hospitals, energy generation, and information systems.

Attitudinal changes within an organization are as crucial as the application of the proper CP/P2 technologies and best practices. A change in attitude on the part of company directors, managers, and employees is crucial to gaining the most from CP/P2 programs, and indeed an EMS works most effectively within the context of a proactive "pollution" management scheme.

CP/P2 is the application of know-how aimed at improving efficiency, adopting better management techniques, improving housekeeping practices, and refining company policies and procedures. Typically, the application of technical know-how results in the optimization of existing processes. There are a number of ways in which technical improvements can be achieved:

- By changing manufacturing processes and technology.
- By changing the nature of process inputs (ingredients, energy sources, recycled water, etc.).
- By changing the final product or developing alternative products.
- By on-site reuse of wastes and by-products.

Common types of CP/P2 options include

- Housekeeping improvements to work practices, safety, and proper maintenance can produce significant benefits. These options are typically low cost or no cost.
- Process optimization improvements.
- Optimizing existing processes can reduce resource consumption. These options are typically low to medium cost.
- Raw material substitution. Some environmental problems can be avoided by replacing hazardous materials with more environmentally benign materials. These options may require changes to process equipment.
- Adopting new technologies can reduce resource consumption and minimize waste generation through improved operating efficiencies. These options are often highly capital intensive, but payback periods can be attractive.
- New product design is an important option. Changing product design can result in benefits throughout the life cycle of the product, including reduced use of hazardous substances, reduced waste disposal, reduced energy consumption, and more-efficient production processes. New product design is a long-term strategy and may require new production equipment, pilot testing, plant trials, and marketing efforts.

When a company invests in a CP/P2 program and technologies, it replaces end-of-pipe pollution control technologies, which add on to the costs of production. It also reduces the risks to a company.

Consider Figure 2–4, which places CP/P2 in relationship to the other available options of managing pollution. To understand this figure better, consider the following. Any waste that is generated must be dealt with in an environmentally effective manner. As an example, Illinois Central Railroad (case study 3 in Chapter 1) disposed of chemically treated railroad ties by incineration in a wood-waste boiler. ICR had a high cost associated with collecting, staging, transporting the waste, and paying for incineration of the wastes. The disposal technology it chose for managing the waste was controversial and environmentally damaging. It created air pollution,

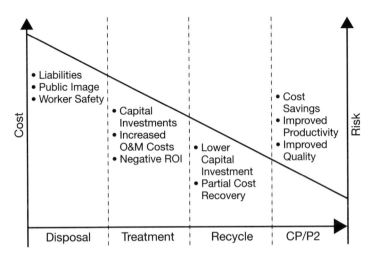

Figure 2-4 The pollution prevention hierarchy.

which exposed a neighboring community to a high health risk. The disposal practice also produced a hazardous ash, which was mismanaged. The ash was spread onto the property of the facility that burned the waste. That ash became airborne and also created a hazardous air pollution problem. The toxic ash was also mobile in storm water and was carried off-site, where the community came into contact with the waste. Illinois Central could have chosen a secure landfill to dispose of its waste, thereby reducing its liabilities and acting in a responsible manner. Illinois Central could have also elected to use concrete railroad ties, which would have prevented the waste from the start. As an example, Rocla Concrete Tie, Inc., is a leading manufacturer of prestressed concrete railroad ties in the United States. This company has manufactured prestressed concrete railroad ties and turnout ties for Class I railroads, commuter passenger operations, transit authorities, and industrial operations nationwide for the past 50 years.[11] Finally, ICR's choice to rely on incineration technology per se was not necessarily bad, but the railroad appears

---

[11]Rocla Concrete Tie, Inc. (Bel Air, MD), Web site: www.roclatie.com.

to have failed in exercising adequate due diligence in selecting a facility that responsibly managed its wastes. With a properly managed EMS, the environmental transformation issues created by the mismanagement of ICR's wastes likely could have been avoided since the EAs would have been identified and proactively addressed.

Whenever a waste is generated it remains a liability forever. Even if that waste is not regulated at the time it is being generated and the practices and technologies of the day are relied upon, there are no guarantees that it will not become a recognized hazard or regulated waste in the future. The less waste generated, the less liability exists for a company. If some of the waste can be recycled or reclaimed, then not only is liability reduced, but some cost recovery may be possible to help pay for the management of the waste. Ideally, if one can eliminate or minimize the waste to begin with, then not only will the cost for treatment, transport, and disposal be reduced or eliminated, but the risks or financial exposures to a company are eliminated. This is not only good business sense, it is most certainly acting in a responsible manner with regard to public safety and the protection of the environment.

When CP/P2 options and pollution controls are carefully evaluated and compared, the CP/P2 options are often more cost effective overall. The initial investment for CP/P2 options and installing pollution control technologies may be similar, but the ongoing costs of pollution control generally are greater than for CP/P2. In addition, the CP/P2 options generate savings through reduced costs for raw materials, energy, waste treatment, and regulatory compliance.

The environmental benefits of CP/P2 can be translated into market opportunities for "greener" products and "green label." Companies that factor environmental considerations into the design stage of a product are generally well placed to benefit from the marketing advantages of any future ecolabeling schemes.

Among the reasons to invest in CP/P2 are these:

- Improvements to product and processes.
- Savings on raw materials and energy, thus reducing production costs.

- Increased competitiveness through the use of new and improved technologies.
- Reduced concerns over environmental legislation.
- Reduced liability associated with the treatment, storage, and disposal of hazardous wastes.
- Improved health, safety, and morale of employees.
- Improved company image.
- Reduced costs of end-of-pipe solutions.

CP/P2 depends only partly on new or alternative technologies. It can also be achieved through improved management techniques, different work practices, and many other "soft" approaches. CP/P2 is as much about attitudes, approaches, and management as it is about technology. CP/P2 approaches are widely and readily available, and methodologies exist for its application.

Because environmental issues are complex, companies are adopting a more systematic approach to environmental management. An EMS provides a company with a decision-making structure and action program to bring CP/P2 into the company's strategy, management, and day-to-day operations. As EMSs have evolved, a need has arisen to standardize their application. A variety of voluntary EMS standards have been developed by industry groups and consensus standards bodies, the best known of which is ISO 14001. In essence, when an EMS is structured around a dedicated CP/P2 program, a company takes a proactive stance against pollution management. In contrast to companies that react to problems, wastes, and out-of-compliance issues on a case-by-case basis, the combination of an EMS and CP/P2 program identifies causes for waste generation, noncompliance, losses, and inefficiencies and focuses on identifying alternatives to systematically reduce these.

The Initial Environmental Review

Overview

The primary objectives of an IER are to obtain the information necessary to progress the program and implement the EMS, identify early savings, and identify areas of noncompliance and liabilities. This is accomplished by

- Reviewing company policies, aims, and objectives in light of the requirements of the program.
- Identifying current liabilities and areas of noncompliance and legal requirements with which the site or company must comply.
- Identifying "early-start" initiatives and producing plans for their implementation.
- Reviewing data collection and management arrangements to identify gaps in and produce recommendations for the improvement of data and monitoring management systems.
- Identifying the information and data requirements necessary to begin quantifying waste, inefficiencies, losses, and pollution.

Once data has been collected and organized, it can be used to help generate the following:

- A baseline of resource use and environmental performance against which future improvements can be measured.
- Benchmarks of the historical and current resource use and environmental performance of the company, site, or processes against typical and good practice.
- An inventory of resource wastage in the company or site, ideally, and where appropriate, in the form of a complete process map.

The IER guides the rest of the program, hence it is crucial that it is objective and carried out by a person(s) with adequate experience and expertise.

An IER provides a broad picture of the environmental aspects within a business's activities and provides the information by which it can begin planning for corrective actions. The primary aims of such an objective review are to obtain the necessary information to progress the EMS as well as to identify early savings and areas of noncompliance and liability.

IER Toolkit

The following checklists and audit questionnaires can be used as a starting basis for identifying key environmental aspects when performing an IER. The reader can modify the questionnaires as appropriate for their facility.

Legal and Due Diligence Compliance Checklist
Aspect Check

PERMITS AND LICENSES

- ☐ For construction or modification of new or existing sites, facilities, and installations
- ☐ For storm-water permit
- ☐ For NPDES (National Priority Discharge Elimination System) permit
- ☐ POTW (publicly owned treatment work) sewer discharge permits
- ☐ For air emissions permits
- ☐ For underground storage tanks (registrations)
- ☐ For Title V permits
- ☐ For RCRA permits
- ☐ For hazard communication and right-to-know training
- ☐ Community emergency planning and response
- ☐ For respiratory fit-testing program
- ☐ For OSHA programs (e.g., lock-out-tag-out, other)
- ☐ For 40-hour hazwopper training

List all areas where documentation is incomplete or under development:

_____
_____
_____
_____

Due diligence monitoring and reporting

- ☐ Hazard communication training records
- ☐ Pulmonary fit testing for respirators
- ☐ Industrial hygiene and medical monitoring records
- ☐ Spill response and contingency plan documentation
- ☐ Preventive maintenance documentation for air pollution controls
- ☐ Preventive maintenance documentation for storm-water treatment equipment
- ☐ Preventive maintenance documentation for wastewater treatment
- ☐ Stack testing documentation
- ☐ Industrial hygiene air-quality monitoring
- ☐ Accident report records
- ☐ Spill incident report records
- ☐ Storm-water monitoring records
- ☐ Odor and other complaints records and company responses

Plans, lists, and photographs

- ☐ Layout plan of the site showing buildings, tanks, open storage facilities, and other significant installations and facilities up to (and including) the site borders. If known, the zones of protected water and the noise sensitivity level of the site and neighborhood should be shown.
- ☐ Map of the vicinity (municipal or district level), showing site location and its proximity to towns, rivers, lakes, etc. (approx. scale 1:1000 to 1:5000)
- ☐ General photos, aerial photos of the site
- ☐ Plans of the gas and sewage systems on the site; details on purpose, age, and materials used
- ☐ Plan of tank facilities and storage capacities above ground and underground with features (e.g., with double walls, protective

structures; and technical data; date of installation or approval and last check)
- ☐ Waste disposal procedures or list of waste and its disposal procedures
- ☐ List of special waste and related disposal procedures
- ☐ Purchasing statistics or inventories of pollutant materials and toxins

EXAMINATIONS, REPORTS, AND CONTRACTS
- ☐ Environmental impact assessment reports
- ☐ Soil analysis reports, laboratory examinations of the soil of the site and off-site
- ☐ Groundwater monitoring data and well registration
- ☐ Materials flow analyses
- ☐ Wastewater analyses
- ☐ Reports on noise measurements
- ☐ Reports on measurements of atmospheric discharges
- ☐ Waste disposal and transporter contracts
- ☐ Copies of correspondence with the authorities (e.g. Environment Protection Agency)
- ☐ Summary reports on incidents
- ☐ Input-output and life-cycle assessments
- ☐ Risk analyses carried out by insurance companies or consultants
- ☐ Copies of incidents recorded by the authorities and other (including those reported internally)

INTERNAL PROCEDURES
- ☐ Internal safety procedures
- ☐ Internal instructions and guidelines etc. on pollutants (e.g., CFCs, asbestos, PCBs)

☐ Documentation on procedures to prevent and accident investigations

IER Audit Questionnaire. Part 1, Water
1.1 Consumption

1. Where does the company's water supply originate? Consider all different sources (e.g., public water supply, groundwater, lakes).
2. How much does your company pay for water per m^3?
3. Is the company's water supply subject to seasonal restrictions due to specific climatic conditions, such as drought, flooding?
4. How much water does the company abstract (m^3 per year)? Differentiate according to the sources.
5. How much water is consumed (m^3 per year)?
6. How is water delivered to the facility?
7. Does the facility monitor the quality of your water supply before it is consumed?
8. Can the facility identify water consumption according to usage (i.e., volume used for processing, cleaning, etc. and the percentage of overall volume this represents)?
9. Is the level of water consumption measured and recorded on a regular basis?
10. Has the facility developed procedures for measuring and recording the volume and quality of water consumed? If so, what are these procedures?
11. What pollutant substances come into contact with water during use (e.g., disinfectants)?
12. Have procedures been developed regarding handling of water pollutant substances? If so, what are these procedures?
13. Does the company reuse process water before discharging it? If so,
    a. What volume of water is reused (m^3 per year)?

b. What percentage of overall water consumption does this reused water represent?
   c. For what is the water reused?
   d. Have you identified the measures that could be implemented to reduce water consumption? If so, were best practices and clean technologies considered when the measures were identified?
   e. Have any of these measures been implemented? If so, what are these measures and have the results of implementing them been recorded and evaluated to see how effective they are?

1.2 Wastewater

1. Does the facility have past records of the volume (m^3 per year) and types of wastewater discharged from your site?
2. Can the facility locate all physical points on the site from which wastewater is discharged?
3. What different types of wastewater are discharged from your site? What volume of wastewater is generated by your site (m^3 per year)? Differentiate according to the different types.
4. What is the origin of each type of wastewater discharged (e.g., cooling, cleaning)?
5. What physical, chemical, and biological properties should be regularly monitored in wastewater?
6. Is wastewater monitored before being discharged?
7. Is the concentration of pollution in wastewater measured and recorded regularly? If so, how often is the concentration of pollution measured?
8. Are the quantities of all pollutant substances detected measured and recorded?
9. What different types of methods and devices for measuring wastewater pollution and quality are used?

10. Are the devices used by the company to measure wastewater pollution and quality checked on a regular basis to ensure that they are working properly?

11. Have procedures been developed regarding
    a. Monitoring and measuring wastewater pollutant content and quality.
    b. Recording these measurements.
    c. Using and checking wastewater measurement devices.
    d. What are these procedures?

12. To where are the different types of wastewater generated discharged?

13. What use is made of the area of water into which the wastewater is discharged (e.g., drinking water supply, agriculture, leisure activities)?

14. Have procedures been developed regarding wastewater discharge? If so, what are these procedures?

15. Does the facility monitor the quality of water in the area into which your wastewater is discharged on a regular basis?

16. What measures, if any, have been identified to reduce the amount of wastewater generated by site activities and reduce or eliminate wastewater pollution?
    a. Which ones were implemented?
    b. Have the results of implementing them been recorded and evaluated to see how effective they are?

1.3 Wastewater Management

1. Does the facility have on-site wastewater treatment facilities? If so, what volume is treated (m^3 per year) and what percentage of overall wastewater does this represent?

2. What treatment process(es) is used for which type(s) of wastewater?

3. Have procedures been developed regarding internal wastewater treatment? If so, what are these procedures?

4. Is treated wastewater reused in your company? If so,
   a. How much wastewater is reused (m^3 per year) and what percentage of the overall volume of wastewater does this represent?
   b. For what is the treated wastewater used?
   c. To where is wastewater discharged once it has been treated?
   d. Does the facility monitor the quality of water in the area into which the treated wastewater is discharged on a regular basis?
5. Does the facility have contracts with external companies to treat or dispose of wastewater? If so,
   a. Do these companies have to meet any environmental requirements?
   b. What are these requirements?
   c. To where is the water transported to for treatment and disposal?
6. Are on-site treatment facilities regularly checked to ensure that they function properly?
7. Are any wastewaters discharged to a public sewer that is connected to a POTW? If so,
   a. How much is discharged per year (m^3)?
   b. What pretreatment is done?
   c. How is the effluent monitored?
   d. Who performs the monitoring?
8. To what are the primary wastewater parameters monitored and controlled? List each effluent parameter and provide ranges of reported and measured values.
9. Is wastewater used for any on-site land application? If so how much per year and for what purpose?
10. Are there any records of accidental water pollution, such as spills or unplanned releases to the sewer or land in the past?
    a. What were the reasons for past accidents?
    b. What were the environmental and human impacts of past accidents?

c. What measures or procedures have been introduced to reduce or eliminate the risk of such accidents happening again?
11. What would be the impact of an accidental discharge of materials used by the company into its water system or directly into the natural environment?
12. What precautionary measures have been taken to isolate wastewater in the case of an accident (e.g., leakage, spillage)?
13. What emergency procedures does the facility have in case of accidental water pollution?

1.4 Costs and Savings
1. Do you know all the costs associated with
   a. Water consumption (e.g., water charges)?
   b. Wastewater discharge?
   c. Internal wastewater treatment (e.g., investment in treatment installations)?
   d. External wastewater treatment (e.g., services provided by external companies)?
2. Has the facility identified potential or realized cost savings from reducing or eliminating water consumption, discharge, or pollution (e.g., reduced water charges)?
3. What are the costs for discharge to the POTW sewer (m^3/yr)?
4. If on-site treatment or pretreatment is used, does the facility know what it is costing per m^3?
5. Prepare a list of the costs and savings already identified.

IER Audit Questionnaire. Part 2, Soil and Groundwater
2.1 Impacts
1. Has a history of operations on the site since they began been compiled to determine possible soil and groundwater pollution?
2. Do you know what substances should be monitored in terms of soil and groundwater pollution?

3. Have any analyses of the soil and groundwater been done to check for pollution below and around your site (either by the company or by an external party)? If so, have the results of such analyses been recorded?
4. Has any pollution been detected in the soil or groundwater below or around your site? If so,
   a. Do you know when (before or after your activities began on the site) this pollution dates from and how it occurred?
   b. Are soil and groundwater analyzed regularly to check for pollution (either internally or by external companies)?
   c. Does the facility know about the different devices for measuring soil and groundwater pollution?
   d. Are devices used by the company to measure soil or groundwater pollution checked regularly to ensure they are working properly?
5. Have procedures been developed regarding
   a. Analysis of the content of soil and groundwater?
   b. Recording the results of these analyses?
   c. Using and checking soil and groundwater pollution measurement devices? If so, what are these procedures?
6. Have all areas on the site where the ground should be made impermeable been identified to avoid soil and groundwater pollution (e.g., chemical storage areas)? If so, have all of these areas been made impermeable?
7. If the facility operates an on-site dump or landfill, has it established a management or remediation plan for this area once it reaches full capacity?

2.2 *Treatment*

1. Has the company already been obliged to remediate or decontaminate polluted soil or groundwater underneath or around the site?
2. Have the results of implementing these measures been recorded and evaluated to see how effective they were?

3. What was remediated and how much?
4. Was a formal site closure placed into effect? If so, for what reason?
5. What were the costs for remediation?

2.3 Accidental Releases

1. Does the facility have records of accidental soil or groundwater pollution in the past?
2. What were the reasons for past accidents?
3. What was the environmental and human impact of past accidents?
4. What measures or procedures have been introduced to reduce or eliminate the risk of such accidents happening again?
5. What emergency procedures do you have in the case of accidental soil or groundwater pollution?
6. Does the facility have a written spill response and prevention plan?

2.4 Costs

1. Do you know the costs associated with preventing soil and groundwater pollution and remediating or decontaminating polluted soil and groundwater?
2. Make a list of the costs you already know about.

IER Audit Questionnaire. Part 3, Atmospheric Emissions and Odors

3.1 Impacts

1. How do the climatic and geographic factors specific to the site influence the atmospheric emissions and odors caused by activities (e.g., exposure to prevailing winds)?
2. Does the facility have past records of the types and quantities of emissions, exhausts, and dust generated by its operations?

3. Has an inventory been done of on-site activities that may cause an odor problem?
4. Does the facility know which air pollutant substances should be monitored?
5. Can the facility locate all points from which atmospheric emissions and odors are emitted?
6. Can the facility identify the origin of these emissions (e.g., materials used during processing)?
7. What types and quantities of atmospheric emissions are generated by the on-site activities? Differentiate according to the source of emission.
8. What effect do these emissions have on employee health and the environment and residents in the immediate vicinity of the site?
9. Are the quantities of atmospheric pollutants emitted regularly measured and recorded?
10. Are solvents susceptible of emitting volatile organic compounds (VOCs) used in products or the manufacturing processes?
11. Does the facility know about the different types of devices that can be used to detect air pollutants?
12. Are the devices used to measure atmospheric emissions checked on a regular basis?
13. Have procedures been developed regarding the following and, if so, what are these procedures:
    a. Monitoring and measurement of atmospheric emissions?
    b. Recording the results of monitoring and measurements?
    c. Using and checking atmospheric measurement devices?
14. How many smokestacks are at the facility?
    a. What pollution controls are used on each stack?
    b. Are all permits in place for controls? Which ones are not permitted?

c. Has stack testing been performed? When and what were the results?
   d. Is continuous stack monitoring performed? If so, what is monitored and how?
15. Has the facility identified measures to reduce or eliminate atmospheric emissions and odors? If so, were best practices and clean technologies considered when measures were identified?
16. Have any of these measures already been implemented? If so, what are these measures and have the results of implementing these measures been recorded and evaluated to see how effective they are?

3.2 Treatment

1. Do you have any on-site facilities for treating exhaust air before it is released into the atmosphere (e.g., dust filters)? If so, what kind of facilities are used for which kinds of emissions?
2. Are each of the controls permitted?
3. When are the dates for permit renewal for each point source control?
4. Do you know about the different types of clean technologies available to treat or reduce air pollution?

3.3 Accidental Releases

1. Has an inventory been done of accidental atmospheric emissions in the past?
2. What were the reasons for past unplanned releases?
3. What were the environmental and human impacts of past accidents?
4. What measures or procedures have been introduced to reduce or eliminate the risk of such accidents happening again?
5. What would be the environmental impact of an accidental release of pollutant substances used in your operations into the air?
6. What emergency procedures do you have in the event of accidental atmospheric pollution?

3.4 Costs

1. Do you know the costs associated with reducing or eliminating atmospheric emissions (e.g., installation of filters)?
2. What are the costs for air permit renewals?
   a. Are these true costs that include management and labor?
   b. Can the hidden costs be broken out?
3. Make a list of the costs the facility already knows about.

IER Audit Questionnaire. Part 4, Noise and Vibrations

4.1 Impacts

1. Has an audit been performed identifying the sources and levels of noise and vibrations under normal operating conditions?
2. What different sources of noise and vibrations are on the site of your activities?
3. Have you had complaints about the noise of your activities from the local residents? If so, what have these complaints been about?
4. Are noise and vibration levels monitored and recorded regularly at the source and limits of your site?
5. Do you know about the different types of devices to measure noise levels and vibrations?
6. Are the devices used by your company to measure noise levels and vibrations checked on a regular basis?
7. Have measures to reduce or eliminate noise levels and vibrations been identified (e.g., soundproofing of premises, stopping deliveries at nighttime)? If so, were best practices considered when the measures were identified?
8. Have any of these measures already been implemented? If so, what are these measures and have the results of implementing them been recorded and evaluated to see how effective they are?
9. What specific engineering controls are there in place for noise control?
10. What specific managerial tools are in place for noise control?
11. What specific personal protective equipment (PPE) is used for hearing protection of workers?

12. Does the company conduct physical examinations for employees to check for hearing loss? How often is this done and are the results published or accessible to employees?
13. Does the company train employees on hearing loss prevention? Are the training records maintained?

4.2 Accidents

1. Does the company have records of accidents that caused the level of noise or vibrations to become unacceptable for employees or local residents?
2. What were the reasons for past accidents?
3. What were the environmental and human impacts of past accidents?
4. What measures or procedures have been introduced to ensure that such accidents do not happen again?
5. What emergency procedures do you have if noise or vibrations reach unacceptable levels?
6. Are employees aware of the time-weighted average (TWA) threshold-limit values (TLVs)?

4.3 Costs

1. Does the company know the costs associated with reducing noise and vibration levels (e.g., installation of soundproof walls)?
2. Make a list of the costs you already know about.

IER Audit Questionnaire. Part 5, Energy

5.1 Consumption

1. Are there past records of the amount of energy consumed by activities?
2. Can you identify the points on your site at which energy is consumed?
3. How much energy is consumed by your operations (kWh per year)?

4. What types and quantities of energy are used by your company? Differentiate between sources (e.g., fuel in m³ per year, gas in kWh per year), renewable and nonrenewable energy, and external supply and own production.
5. How much energy is consumed per unit of production and per employee?
6. What is the level of energy consumption for each source as a percentage of overall energy consumption (e.g. 75% natural gas, 25% electricity)?
7. Can you clearly show your energy consumption according to usage (e.g., the amount used for processing, heating) and the percentage this represents of overall energy consumption?
8. Is the level of energy consumption measured and recorded on a regular basis?
9. Do you operate heat recovery facilities (e.g., heat recovery from an incineration unit)? If so,
    a. How much heat is recovered?
    b. What percentage of total consumption does this represent?
    c. For what is the recovered heat used?
10. Have measures been identified to reduce energy consumption? If so, were best practices and clean technologies considered when these measures were identified?
11. Have any of these measures been implemented? If so, what are these measures and have the results of implementing them been recorded and evaluated to see how effective they are?
12. What are the various fuel sources used to supply energy? Can you develop yearly estimates for each source?

5.2 Impacts

1. What atmospheric emissions are generated by your energy consumption (e.g., CO_2)?
2. Are these emissions measured and recorded on a regular basis?
3. Do you know about the different types of devices that can be used to measure emissions generated by your energy consumption?

4. Are the devices used by your company to measure emissions generated by energy consumption checked on a regular basis?

5. Have regulatory problems been associated with any sources of energy use? If so, what?

5.3 Accidents

1. Do you have a record of any accidents related to your energy facilities (e.g., a broken thermostat leading to overheating and explosion of a boiler)?

2. What were the reasons for past accidents?

3. What were the environmental and human impacts of past accidents?

4. What measures or procedures have been introduced to ensure that such accidents do not happen again?

5. What emergency procedures do you have in the case of accidents relating to your energy facilities?

5.4 Costs and Savings

1. How much does your company pay per year for its energy consumption?

2. Can you trace the changes in energy bills over time and the reasons for changes?

3. Do you know the costs associated with reducing energy consumption (e.g., installing a new heating system)?

4. Have you identified any potential or realized cost-savings from energy efficiency?

5. Make a list of the costs and savings you already know about.

IER Audit Questionnaire. Part 6, Waste

6.1 Waste Generation

1. Does the company have past records of the amounts and types of waste generated by the operations?

2. What types of waste are currently generated by on-site activities?

3. Is the volume of waste generated measured and recorded regularly?
4. How much solid waste is generated (tonnes per year)? Differentiate according to the types of waste.
5. Can any of the wastes generated by activities be defined as hazardous? If so, which ones and why?
6. Is the facility permitted as a large quantity, small quantity, or conditionally exempt generator under RCRA legislation (for U.S. facilities)?
7. How much hazardous waste is generated by the company (tonnes per year)? Differentiate according to the types of waste.
8. What are the sources of the hazardous waste generated?
9. Does the facility recycle or reuse any of its wastes internally? If so,
    a. What quantities and types of waste are reused?
    b. What percentage of overall waste does this represent?
    c. For what are the waste materials reused?
10. Have any measures been identified for reducing or eliminating waste or recycling waste? If so, were best practices and clean technologies considered when these measures were identified?
11. Have any of these measures been implemented? If so, what are these measures and have the results of implementing them been recorded and evaluated to see how effective they are?

6.2 Handling and Storage

1. How are the different wastes generated by the facility's operations collected and stored?
2. Can you clearly identify all points on your site where wastes are collected and stored?
3. Are the contents of the storage containers clearly labeled?
4. Are storage facilities inspected regularly to ensure they are intact and correctly labeled?

5. What procedures have been introduced for the collection and storage of the different wastes?
6. What procedures and instructions have been introduced for handling hazardous wastes?
7. Are there special storage conditions for hazardous waste before it is disposed of?
8. Does the facility use any kind of electronic waste tracking system to manage stockpiling and disposal of wastes?
9. Are drums stored on pallets?
10. Are there written procedures for managing corroded and leaking drums?
11. How often are waste storage locations inspected for rusted and leaking drums?
12. Are there any air emissions problems associated with on-site waste stockpiles (e.g., airborne dust)?

6.3 Treatment

1. Are there any internal treatment or pretreatment facilities for your waste materials? If so, what processes are used?
2. Do you know about the different methods for treating the types of waste generated by your company?
3. Are any materials recovered during treatment or pretreatment? If so,
   a. Which materials are recovered and in what quantities?
   b. For what are these materials used?
4. Have procedures been introduced regarding treatment of different types of waste? If so, what are these procedures?
5. Does the facility have contracts with external waste treatment companies? If so, do these companies have to meet specific environmental requirements?

6.4 Disposal

1. How are the different types of waste sorted before being disposed of or treated?

2. Are there appropriate disposal channels for each type of waste?
3. Where does waste end up once it has left the site of your operations?
4. Are any of the waste materials recycled or reused externally?
5. Do you return any waste directly to the supplier? If so, do you know if the supplier recycles or reuses this waste?
6. Have procedures been introduced regarding disposal of different kinds of waste? If so, what are these procedures?
7. Are records kept of hazardous waste disposal (e.g., consignment tracking numbers)?
8. Are hazardous wastes correctly sealed and labeled for transport?
9. Is any treatment or stabilization of wastes practiced? If so, describe.
10. Are any of the wastes ignitable?
11. Are any of the wastes flammable?
12. Are any of the wastes corrosive?
13. Are any of the wastes toxic?
14. Are any of the wastes classified as forming leachates?
15. Does the facility track waste reduction performance over time?

6.5 Accidents

1. Does the facility maintain records of past accidents during waste handling, storage, treatment, or disposal?
2. What were the reasons for past accidents?
3. What were the environmental and human impacts of past accidents?
4. What measures or procedures have been introduced to ensure that such accidents do not happen again?
5. What emergency procedures do you have in the event of accidents during waste handling, storage, treatment, or disposal?

6.6 Costs and Savings

1. Do you know all the costs associated with
   a. Waste disposal (e.g., collection fees)?
   b. Internal waste treatment or pretreatment (e.g., investment in treatment facilities)?
   c. External waste treatment (e.g., fees paid to waste treatment companies)?
2. Have you identified any potential or realized cost savings from reducing, eliminating, reusing, or recycling waste (e.g., reduction of waste disposal fees)?
3. Make a list of the costs and savings you already know about.

IER Audit Questionnaire. Part 7, Raw Materials
7.1 Raw and Operating Materials

1. Are there past inventories of the quantities and types of raw and operating materials used by the facility?
2. What quantities and types of raw and operating materials are currently used?
3. Are records of the cost and origin of these materials maintained?
4. What materials make up the raw and operating materials you use?
5. Do you have a coding or classification system for the different types of materials?
6. If you use materials that contain hazardous substances, are they clearly labeled?
7. Do you have a register of hazardous materials bought, stored, processed, and transported by your company?
8. What are the environmental impacts of producing your raw and operating materials?
9. Do you have guidelines for purchasing raw and auxiliary materials? If so, what environmental criteria do they include (e.g., buy biodegradable cleaning products)?
10. Do suppliers of your materials have to meet specific environmental requirements? If so, what are these requirements?

11. Can you follow the path of hazardous materials used by your company, from the time when purchased to when they are discharged from your company as waste?
12. Have you identified measures for reducing or eliminating the amount of materials used? If so, what are these measures?
13. Are storage areas for raw and operating materials clearly labeled and equipped, depending on the types of materials stored (e.g., with fire protection devices)?
14. Is access to storage areas of hazardous materials regulated?
15. Have you already implemented any of these measures? If so, what are these measures and have the results of implementing them been recorded and evaluated to see how effective they are?

7.2 Semifinished Goods and Office Supplies

1. Does the company maintain records of the type, quantity, cost, and source of semifinished goods and office supplies purchased?
2. Do you have any guidelines for purchasing such goods? If so, what environmental criteria do they include (e.g., always buy recycled paper)?
3. Do suppliers of these goods have to meet any environmental requirements? If so, what are these requirements?

7.3 Packaging Materials

1. What quantities and types of packaging are used during delivery and storage of raw and operating materials, semifinished goods, and office supplies?
2. Do these materials contain any toxic or hazardous substances?
3. Do you apply environmental criteria when deciding which packaging materials to use? If so, what are these criteria?
4. Have you identified any measures for reducing or eliminating the amount of packaging used for delivery and storage of materials, semifinished goods, and office supplies?

5. Have any of these measures already been implemented? If so, what are these measures and have the results of implementing them been recorded and evaluated to see how effective they are?

7.4 Costs and Savings

1. Do you know the costs associated with reducing or eliminating usage of materials, semifinished goods, and office supplies?
2. Have you identified any potential or realized cost savings from measures to reduce, eliminate, or substitute the materials, semifinished goods, and office supplies you use?
3. Make a list of the costs and savings you already know about.

IER Audit Questionnaire. Part 8, Products

8.1 Design

1. Were any environmental criteria applied during the design of your existing products? If so, what are these criteria?
2. Are environmental criteria applied when new products are designed (e.g., that products should be reusable, recyclable, easily disassembled)? If so, what are these criteria?
3. Are employees familiar with or have been trained on design for environment or life-cycle tools?

8.2 Packaging

1. Has an inventory been done of the quantities and types of packaging used during storage, packaging, and transport of your finished products?
2. What quantities and types of packaging currently are used for your finished products?
3. Do the packaging materials you use contain any toxic or hazardous substances?
4. Do you apply any environmental criteria when choosing packaging materials? If so, what are these criteria?

5. What percentage of the packaging used for your products is reusable or recyclable?
6. Can purchasers of your products return the packaging to your company?

8.3 Usage

1. Do you know the environmental impacts of your products when the final product is being used?
2. Do you provide customers with information on minimizing the environmental impact during usage?

8.4 Disposal

1. Where are your products and their packaging disposed of once they have been used?
2. Do you provide users of your products with instructions for disposal?
3. Can users of your products return them to you at the end of the product life?
4. Does your company have the capacity to recycle or reuse all or parts of returned products or their packaging? If so, which parts are reused and for what purposes?

8.5 Costs and Savings

1. Do you know the costs associated with
   a. Designing products to reduce their environmental impact?
   b. Reducing, eliminating, and substituting materials used during manufacture?
   c. Reducing, eliminating, and substituting materials used to package your products?
   d. Collecting, recycling, and reusing your products at the end of their life?
2. Have you identified potential and realized cost savings from
   a. Designing products to reduce their environmental impact?

b. Reducing, eliminating, and substituting materials used during manufacture?
   c. Reducing, eliminating, and substituting materials used to package your products?
   d. Collecting, recycling, and reusing your products at the end of their life?
3. Make a list of the costs and savings you already know about.

IER Audit Questionnaire. Part 9, Logistics

9.1 Impacts

1. What different types of transport are used by your company and for what purposes (e.g., heavy vehicles for delivery of raw materials, rail for distribution of finished goods)?
2. Do you know the fuel efficiency and emission levels of the vehicles used by your company?
3. If you use external delivery or distribution companies, do you consider the environmental impact of their services?
4. Have you identified measures for reducing the environmental impact of transporting your goods (e.g., using other types of transport, conversion to low-emission vehicles)?
5. Have any of these measures been implemented? If so, what are these measures and have the results of implementing them been recorded and evaluated to see how effective they are?
6. Is water used for vehicle washing and maintenance? How much? Are any water efficient technologies and practices being used?

9.2 Costs and Savings

1. Do you know the costs associated with reducing or eliminating the environmental impacts of your logistics (e.g., cost of purchasing more fuel-efficient delivery trucks)?
2. Have you identified potential or realized cost savings from measures to reduce or eliminate the environmental impact of your logistics?
3. Make a list of the costs and savings you already know about.

3
Integrating Cleaner Production with an EMS

Introduction

A policy commitment to pollution prevention (P2) or cleaner production (CP) is among the highest priorities the ISO 14001 EMS standard requires. This is because, within the spirit of any EMS, it is recognized that end-of-pipe treatment technologies, practices, and approaches to dealing with pollution, waste, and inefficiencies tend

- To alter only the medium, method, and location of waste disposal.
- To generally consume even more energy and other resources in the cleanup and waste disposal process.
- To add to the costs of production.

Moreover, unless a waste is completely immobilized or destroyed, it remains a liability forever. It makes little difference whether the waste is a recognized hazardous material that is regulated today or not. Recall some of the litigation case studies from Chapter 1. Companies like Lockheed-Martin certainly can argue that, in 1956 when the Redlands facility started up, there were no legally enforceable environmental standards that guided the company on the waste disposal practices and technologies to rely on for managing waste ammonium perchlorate (a rocket propellant fuel ingredient) or even TCE for that matter. But, decades later when health effects were identified, the company was not only liable the costs for contaminated groundwater remediation, it faced a class action toxic tort.

By contrast P2, as ISO 14001 notes in its definitions section, tends to reduce adverse environmental impacts while improving production efficiency and reducing costs; and because pollution is reduced or eliminated at the source, liabilities are substantially reduced. Indeed, arguably, most of the environmental and economic performance benefits, reduction in risk of environmental liability, and hence the sustainability benefits of an EMS derive from the P2 activities and investments that it generates. On an even broader scale, cleaner production is simply smart business sense, because the overall focus is not on generalized forms waste or pollution but implies high efficiency. The more efficient a business is, the higher its productivity and hence the more profitable it is.

In short, there is no such thing as a meaningful EMS without P2 or CP at its core. Despite this, EMS guidance documents most often fail to provide instruction on how to identify and evaluate P2 opportunities in an enterprise. And, indeed, I visited a number of facilities that even have formal ISO 14001 certification yet P2/CP programs are not practiced on a dedicated basis. In such cases, the EMS itself operates at low efficiencies, thereby explaining why these companies remain threatened by competitors, are not optimizing their production costs, and continue to be plagued by noncompliance issues.

This chapter is adapted from an earlier publication in which the scheme for an integrated EMS/P2 program was explained.[1]

---

[1] A. Bendavid-Val and N. P. Cheremisinoff, *Achieving Environmental Excellence: Integrating P2 and EMS for Improved Profitability* (Washington, DC: Government Institutes, 2003).

When the EMS is fully integrated with CP/P2 activities, the EMS is implemented faster and operates more efficiently.

The Link between Pollution Prevention and an EMS

P2/CP is not just a concept, an idea, an approach, or a mindset. Rather, it requires detailed analysis and planning that incorporates a rigorous methodology and a set of engineering, management, and life-cycle costing tools. True, you can walk around the shop floor and identify many P2 housekeeping and relatively low-cost investment opportunities—leaking valves, lights left on unnecessarily, heat losses from poor insulation, materials that could be recycled internally, waste that could be reduced with automated controls, ineffective steam traps, and so on. But to identify the bigger-ticket P2 opportunities that yield major production process efficiencies and cost savings, properly evaluate those opportunities from a bottom-line financial point of view, as one would do with any investment plan, and then comparatively evaluate them and devise a strategy, and to do this on a continual improvement basis, requires a well thought-out program. The structural elements of such a program are illustrated Figure 3–1.

The P2/CP program begins with three steps that need to be taken as part of setting it up the first time:

Step 1. Organizing to put the program in place.

Step 2. Gathering and interpreting basic production process information.

Step 3. Confirming top management's commitment to a program of continual improvement in preventing pollution and implementing cleaner production technologies.

Once the program is implemented, the focus is on maintaining a continual improvement program. Combined, these steps amount to an adaptation of the famous plan-do-check-act management system model.

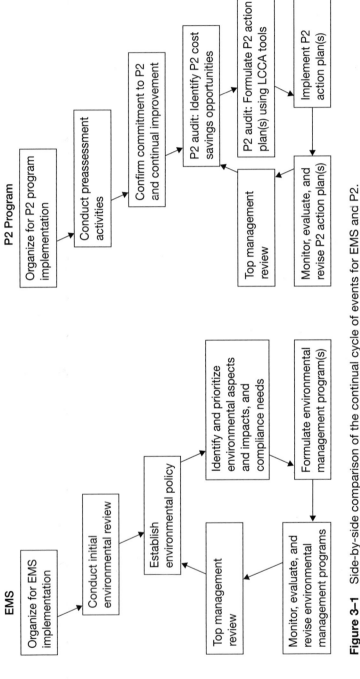

Figure 3–1 Side-by-side comparison of the continual cycle of events for EMS and P2.

Source: Reproduced with the permission of the publisher from A. Bendavid-Val and N. P. Cheremisinoff, *Achieving Environmental Excellence: Integrating P2 and EMS for Improved Profitability* (Washington, DC: Government Institutes, 2003).

The necessarily similar structures of EMSs and P2/CP programs make it possible to integrate the elements of a P2 program smoothly into an EMS. When this is done, P2/CP is raised from being a mere commitment within the EMS to being a major operational component of the EMS.

- Instead of looking just at the regulatory compliance status and environmental aspects of your company's operations then at ways to improve on these, you would also be specifically and systematically looking for high-return P2 opportunities from the outset and doing so continuously.

- An integrated EMS/P2 program reduces operating costs and the risks of environmental liability more, better, and faster than a traditional EMS. The earlier major cost savings can be used to help finance investments, further accelerating improvements in business and environmental performance for the company.

Let us now focus on how an integrated EMS/P2 program can work.

Integrating EMS and CP/P2

The integrated EMS-CP/P2 model is conceptually presented in Figure 3–2. In this approach, a commitment and resources are placed under a CP/P2 tent and moved to the front of the EMS implementation process and the EMS continual improvement cycle. In addition, the integrated model has some features that make it more responsive than traditional EMS models, like ISO 14001, to the meet the heightened concern that companies in the post-9/11 era have with environmental liability, security, and emergencies and to the growing interest in corporate responsibility and constructive communication with stakeholders. Readers familiar with ISO 14001 will recognize that the overall flow of the integrated EMS/P2 model in Figure 3–2 is the same, and yet not the same, as in a typical EMS.

As with a traditional EMS, the first step in an integrated EMS-CP/P2 program is to establish commitment, responsibilities, awareness,

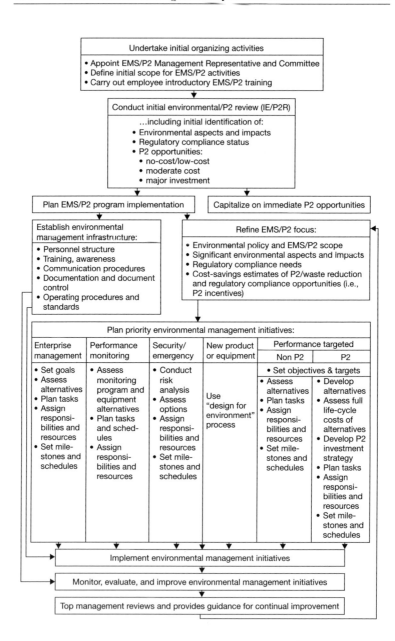

Figure 3–2 The integrated EMS/P2 model.

Source: Reproduced with the permission of the publisher from A. Bendavid-Val and N. P. Cheremisinoff, *Achieving Environmental Excellence: Integrating P2 and EMS for Improved Profitability* (Washington, DC: Government Institutes, 2003).

and an initial scope. This element of the program model in Figure 3–2 shows the three most critical of the many possible initial organizing activities that a company might undertake.

The initial environmental and P2 review is the EMS-CP/P2 program's counterpart to the EMS-related initial environmental review. A company may conduct an IER to establish a baseline on where it stands with regard to its environmental compliance and performance and to develop related information helpful for planning the steps to put its EMS in place. Typically, an IER does not develop information that would help in planning a P2 program to any significant degree. That is because an EMS, at least one similar to ISO 14001, starts with the company's environmental interactions ("aspects"), so the IER is oriented to that. In contrast, an IER that incorporates a CP/P2 program, apart from housekeeping, starts with inefficiency and waste in the company's production processes, and so its initial data collection would need to be oriented to that. This can be referred to as an IE/P2 review (IE/P2R).

Examples of some of the P2-specific information one would collect during an IE/P2R that normally would not be collected during a traditional IER include

- Details of unit processes.
- An initial list of unit operations and their functions within those processes.
- Basic process flow sheets.
- Basic data for materials and energy balances.
- An initial assessment of causes and sources of significant pollution, waste, low productivity, poor quality, and energy losses along the manufacturing trains in the company.

Through an IER that focuses on P2 opportunities, a company can identify losses and cost incentives for corrective actions that are much more numerous, significant, process based, and data based than would be the case through a traditional IER that focuses largely on compliance-related issues; and as a result of the IE/P2R, it will also have process descriptions and baselines in hand for use during the EMS/P2 program.

- A company should capitalize on immediate P2 opportunities. Why wait? Since, as a result of the IE/P2R, the company will have already identified some solid P2 opportunities at this point, it makes sense to start reducing costs and improving environmental performance right away, even before the EMS/P2 program has been implemented. Of course, jumping in this way, your company has to be careful to invest only in P2 opportunities with clearly high rates of return, probably mostly in the no-cost/low-cost and moderate-cost categories. Later on in the EMS/P2 program, the company will develop a carefully considered P2 investment strategy. But capitalizing on immediate P2 opportunities can provide powerful education and motivation among employees for its EMS/P2 program, and the cost savings can even help offset the costs of the EMS/P2 program.

- Early planning of the EMS/CP-P2 program implementation can be performed at the same time. While undertaking its first concrete P2 actions and investments, the company is planning the steps to implement its permanent EMS/P2 continual improvement process. At the same time, as shown by the arrow to the vertical box on the lower-left side of Figure 3–2, the company is planning the steps to establish its environmental management infrastructure.

- The team should be prepared to refine the EMS/P2 focus. Here, the company refines the initial determinations made during the IE/P2R. The left side of this box in Figure 3–2 is roughly equivalent to clauses 4.2 through 4.3.2 of ISO 14001, covering environmental policy, aspects, and legal requirements. In the EMS/P2 program model, in addition, you would extend, deepen, and refine the estimates of regulatory compliance and production process costs, potential environmental liabilities, and the related P2 cost savings and risk reduction based on available technologies. These together would provide the basis for setting goals, objectives, and targets for the company's environmental management initiatives. Again, the combination of a traditional EMS "environmental aspects" approach with the production process approach of a P2 program yields a much more powerful set of goals, objectives, and targets than either an EMS or a P2 program alone would yield. The CP-P2 analysis here would involve certain steps from phase 2 of the P2 audit model (see Chapter 5).

- The company will need to establish the environmental management infrastructure. Environmental management infrastructure is organizational infrastructure meant to support sound environmental practices on a day-to-day basis as well as support specific environmental management initiatives generated through the EMS/P2 program. This box in Figure 3–2 is roughly equivalent to clause 4.4 (implementation and operation) of the ISO 14001 EMS standard.

The model in Figure 3–2 reflects that environmental security and emergency preparedness should be a type of environmental management initiative rather than a part of the supporting infrastructure. This would mean that, under the EMS/P2 program, the company always is looking for initiatives that would help it improve in this area, and therefore environmental security and emergency concerns receive more deliberate, continuous, and systematic scrutiny.

- Note the plan priority environmental management initiatives box in Figure 3–2. This is the counterpart of ISO 14001's clauses 4.3.3, objectives and targets, and 4.3.4, environmental management programs. By *environmental management initiatives*, we mean carefully planned activities designed to address the company's needs for improvement that have been identified, refined, and prioritized in earlier phases of EMS/P2 program work. These initiatives are basically the same as what ISO 14001 calls *environmental management programs*, and what the *Green Profits*[2]

---

[2] N. P. Cheremisinoff and A. Bendavid-Val, *Green Profits: A Manager's Handbook to ISO 14001 and Pollution Prevention* (Oxford: Butterworth-Heinemann Publishers, 2001), pp. 373–316.

P2 audit model calls *action plans*. The prominence of this box in Figure 3–2 is meant to emphasize that environmental management initiatives are the centerpiece of an EMS/P2 program.

In Figure 3–2, we show six possible types of environmental management initiatives. Each type of initiative requires its own planning process, as suggested in shorthand in the column for each type. The types, or categories, of environmental management initiatives reflect the areas of environmental management concern the company's management wants to be sure are the primary focus of its continual improvement efforts. We do not have space in this book to dwell on what is encompassed by each type of environmental management initiative shown in Figure 3–2 and its unique planning requirements, much as we would like to.

But, we do need to look for a moment at the types of initiatives shown in the two columns at the right of this box. These are environmental management initiatives aimed at achieving specific performance targets, incidentally the only kind of environmental management program explicitly required in ISO 14001 and most other EMS models. As the illustration suggests, you can employ a fairly routine planning process when for whatever reason a non-P2 approach must be used to achieve a particular performance target. For the collection of performance targets that will be achieved through P2 techniques and technologies (the rightmost column), you need to perform extensive economic analysis, including life-cycle costing, and develop a full P2 investment strategy. The investment strategy may even be structured so that the economic benefits of earlier P2 investments help finance subsequent P2 investments. In any case, the P2 planning work at this point would involve many of the steps in phases 2 and 3 of the CP/P2 audit model presented in *Green Profits* and outlined in Chapter 5.

Planning for EMS/P2 environmental management initiatives is followed, as shown in Figure 3–2, by the familiar steps of implementation; monitoring, evaluation, and revision; top management review; and the next cycle of continual improvement. As with ISO 14001, once the pieces are in place, the integrated EMS/P2 program does not really operate as a cycle with discrete sequential elements, rather, all its elements should be operating more or less all the time.

The integrated EMS/P2 program described here provides a framework that accommodates all the requirements of ISO 14001 and, therefore, can be used to pursue ISO 14001 certification, if that is what your company wishes to do. But an integrated EMS/P2 program can yield environmental and economic performance benefits faster than typical ISO 14001–type EMSs, and the model offers other advantages as well.

The Role of an Environmental Management Information System

The reader can appreciate that substantial pressures are placed on companies to act more responsibly for care of the environment, and as illustrated by the case studies in Chapter 1, exercising responsible care for public safety and the quality of life are integral if not the major thrust behind being able to ensure good environmental performance. In fact, today's businesses simply cannot afford not to make environmental management a major component of their primary business strategies. Contributing factors that strengthen this viewpoint include the globalization of the marketplace, pressures from nonprofit organizations, local legislation and global accords, a burgeoning green supply chain, and a well-educated and informed public, just to name a few.

And, as the reader likely has grasped by now, an EMS like ISO 14001 requires the control and management of a vast amount of information not all of which has direct impact on compliance but does impact on the efficiency of the EMS. Managing such a system manually proves inefficient and could substantially limit the performance of the EMS. This is particularly true for large, complex facilities like refineries. Manual files and data manipulation, archiving, and making efficient use of the vast amount of information having an impact on the environmental action plans, policies, and infrastructure investments that fall out of an EMS can be overwhelming and may never meet the ever-growing information demands of environmental managers. As such, environmental professionals must turn their focus to computerized solutions that enable them to attain their information-intensive management objectives.

An environmental management informational (EMIS) or intelligent system is a vital tool in a forward-thinking environmental manager's toolkit. It is an important component of environmental management that can assist both environmental managers and decision makers fulfill their daily tasks.

An EMIS is a computer-based technology that supports environmental management systems. Tasks that EMISs support include tracking activities, tracking waste, monitoring emissions, scheduling tasks, coordinating permits and documentation, managing material safety data sheets (MSDSs), conducting cost/benefit analysis, and choosing alternative materials, to name a few. Many computer-based tools available on the market today designed to support these types of tasks are labeled EMISs. However, they typically are unconnected and uncoordinated, stand-alone tools implemented in an ad hoc manner. This has created fragmented, confusing, and sometimes conflicting perceptions of EMISs, resulting in the selection of any software package. Ideally, an EMIS should be viewed more holistically. An EMIS should enhance the purpose of an EMS considerably. It should provide a medium that crosses the boundaries of logistics, data, motivation, language, culture, and intellect.

To illustrate how an EMIS works to support an EMS, consider the elements that environmental managers need to take into consideration to successfully maintain and operate an EMS. These elements are illustrated by categories shown in Figure 3–3. When one considers a company with many plants operating around the world, it is a challenge to manage the entire business let alone keep communications between different business units and manufacturing facilities up to date. EMS implementers are faced with managing a changing landscape of environmental issues and must do so economically. They must deal with many system inputs, laws and regulations, aspects, impacts, procedures, stakeholder needs, and the like.

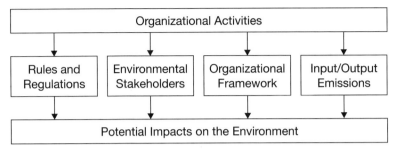

Figure 3–3 How organizational activities have an impact on the environment.

Environmental program management for a business like this requires using office tools that have been set up according to preestablished corporate templates. Often, office tools the support systems uses are not adequate to link geographically separated facilities in ways that enable managers and decision makers to adequately share the information managed by each of their sites. This means that access to information is often limited to the site itself. This makes it difficult for personnel at corporate headquarters to gain access, keep track of performance, and allocate resources effectively.

The success of an EMS depends heavily on the ability to smoothly circulate information from, and to, executive management, middle management, and operations. For the types of information flows, refer to Figure 3–4. While this obstacle can be partially overcome by having the environmental managers at each plant communicate directly with company leaders at the head office on a regular basis, there are more effective tools to streamline communications and enable decisions in a timely manner. In reality, companies need to be able to evaluate their environmental performance on a continual, real-time basis. They also must be able to quickly identify new ways to do business in order to improve performance while reducing potential environmental risks. By carrying out these objectives, companies will be able to recognize how new environmental targets should be defined and supported with the proper resources. They

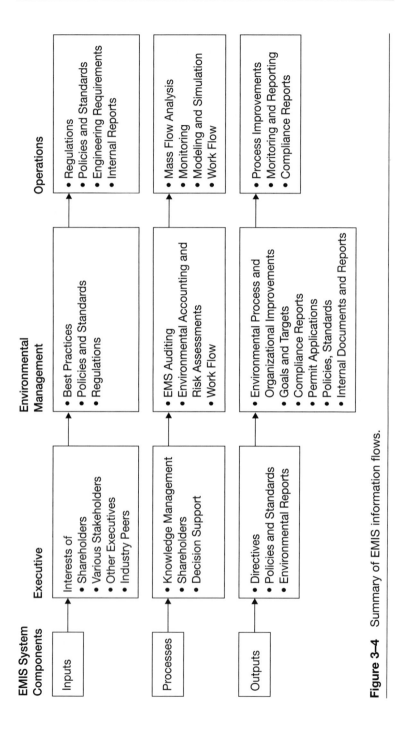

Figure 3-4 Summary of EMIS information flows.

will be able to identify and create appropriate implementation and training plans that can be set up and provided to employees quickly, to provide follow-ups on established performance processes, and to implement audits more rapidly to identify out-of-compliance issues. Such activities should make performance evaluations by managers and decisions to improve the system and monitor performance more efficient.

To manage the environmental aspects of multiple sites and business units on a real-time basis requires the application of cutting-edge technologies that can streamline the *process* of environmental management while not compromising *performance*. Environmental leaders at the corporate level as well as at each geographic location need to apply a strategically-oriented technical approach based on a sound environmental information system framework. Such an approach will assist all stakeholders from within the company to

- Understand EMS language and its requirements.
- Identify and evaluate the environmental risks associated with each site operation.
- Objectively determine the significance of each environmental aspect to prioritize the actions necessary in the environmental management programs (EMP).
- Understand, and stay ahead of, the various laws and regulations that apply and determine the measures that must be taken to comply with them.
- Conduct a gap analysis against the EMS to rate certification readiness to ISO 14001.
- Distribute procedures to all employees while ensuring security of the information and the users.
- Plan all the environmental tasks while enabling follow-ups by employees, managers, or corporate headquarters.
- Monitor the progress of environmental programs at all times.
- Manage audits, nonconformances, and corrective and preventive actions.

- Access the results of previous environmental audits and performance results.
- Share common information from one site to another.
- Provide an outreach to local community and global stakeholders.
- Integrate environmental, quality, and health and safety systems.
- Support uniform training requirements for all the preceding.

On this lengthy list are essential components that, when addressed through an information system, enable a company to remain flexible enough to adapt and evolve with changing environmental, economic, and market conditions.

At the operational level, numerous tools support activities that fulfill compliance and environmental programmatic responsibilities. These tools are usually homegrown, unconnected, and if off the shelf, are specific to tasks. Middle management has more defined information requirements and responsibilities, that is, directly supporting a specific regulation or waste management and site remediation needs in the field. The tools in use at this level focus more on data management and knowledge, like environmental management intranets, environmental auditing assessment software, and MSDS software. Finally, at the executive level, the information requirements are often not defined. There is a high degree of information and data summarization as information and data is reported upwards, that is, environmental auditing information and data on targets and objectives.

Figure 3–5 summarizes several EMIS implementation strategies that have been deployed by companies. The figure demonstrates typical strategies and the benefits, drawbacks, and resources needed to develop it. Generally, when deciding on an information technology approach, companies can apply three EMIS design options on which to base costing scenarios. This additional level of consideration can further optimize the EMIS choice.

The context of any EMIS is as important as the type of EMIS itself. The EMIS design options are summarized in Table 3–1. The Recommended Resources list works where the reader may obtain additional information.

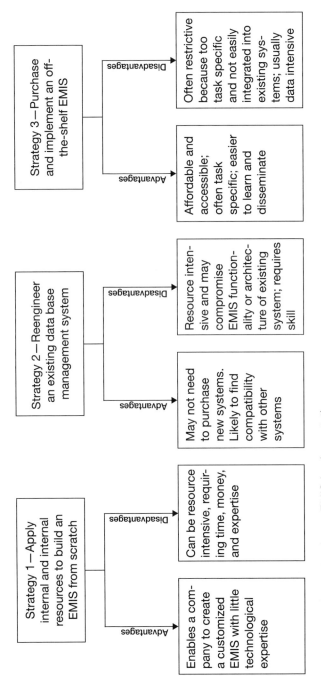

Figure 3-5 Strategies for an EMIS implementation.

Table 3-1 EMIS Design Options

Design Option	Details
Metainformation system	An index of information is available on the location of environmental and business data available in the company.
Virtual database system	Users access and update information via a uniform and standardized Web or client-server interface.
Central database system or data warehousing approach	Environmental data are collected by the various information systems of the company and centralized in the data warehouse so that it can be easily accessed on demand.

Recommended References

Fielding, Stanley. November 2000. "ISO 14001 Brings Change and Delivers Profits," *Quality Digest* 3, no 12.

———. December 1998. *The Environmental Management Report.* New York: McGraw-Hill Book Companies.

Gunther, O. 1998. *Environmental Information Systems.* Berlin: Springer Publishing.

MacLean, Paul, Mariela Tovar, and J. D. Phyper. October 2000. *Finding the Right Solution to Human Performance Problems in Environmental Management,* available at http://www.eem.ca/francais/publications/ISO14001/text-iso-161000-1.html.

4
Improving Competitiveness through an EMS

Introduction

All too often, an EMS is applied with a primary focus given to improved compliance. However, with CP/P2 at its core, an EMS enhances operational efficiency, improves product quality especially when "quality teams" interact with the "EMS/P2 management" team, and captures financial savings as well as environmental credits. But, even there, the potential benefits derived from a fully integrated EMS/P2 program stops short. While a fully functional EMS with P2 at its core can improve the bottom-line economic and environmental performances of a company, it may also serve as a basis for improving performances along the entire supply chain. This can help add to sector robustness and improve world markets.

Benefits That Improve Competitiveness

It costs a company time, staff energy, and money to put an EMS in place and maintain it forever as part of its overall management system. What motivates top management to undertake the effort? In simplest terms, massive historical evidence in the form of case studies, testimonials, and corporate annual and environmental reports show well-run EMSs pay back the investment in them handsomely in terms of dollar cost savings and risk reduction, not to mention other benefits of improved environmental performance and regulatory compliance.

After all, fundamentally, pollution is waste: it represents inputs for which money was paid that leave the plant in a form that is not part of the company's product—essentially, it is money discarded. To make matters even worse from a business point of view, to meet regulatory standards, companies have to pay even more money to capture the pollutants, still more money to process them somehow, and yet more money to dispose of them somewhere else, where they may become a huge financial liability some time in the future. In simplest terms, an EMS pays for itself by providing a framework for systematically identifying and seizing opportunities to *prevent* pollution at the source, to reduce waste (through process changes, technology changes, materials substitution, or other means), and thereby improve operational efficiency and reduce the risk of financial liability. That certainly should be enough of a driver for any competent senior manager.

But to elaborate a bit, because of the characteristics of a sound continual-improvement-based EMS and the direct economic and environmental improvements it generates, establishing and maintaining an EMS in an organization responds to at least 10 types of drivers:

1. *Minimize the cost of operations required for environmental management and compliance.* Because an EMS is built on a widely accepted management principle, it integrates easily with quality management systems, occupational health and safety management systems, and the overall management system of the company, so that a minimum of added cost is required. In fact, many companies have unified environment, health, and safety (EHS) management systems.

2. *Improve environmental performance at lower cost.* Because it builds capacity within the company for strategic environmental

management, an EMS enables a company to be proactive in improving its environmental performance and compliance. Strategic environmental management includes accounting for likely future regulatory changes and continually identifying pollution prevention (P2) opportunities, an approach that always allows for lower-cost solutions than scrambling to meet new regulatory standards when they come into force.

3. *Improve overall enterprise management and overall business performance.* An EMS promotes good business management practices overall, because, through an EMS, environmental responsibility must be clearly allocated throughout the organization, monitoring systems must be established, adequate operational controls must be ensured, proper documentation must be created and controlled, waste and energy consumption must be minimized, and other elements of sound enterprise management must be established and maintained. As an example, many managers of companies in transition economies, like those in Central and Eastern Europe, report that implementing an EMS helped them rationalize their overall management frameworks for the first time.

4. *Prepare for the coming trends in environmental regulation.* There is a gradual movement throughout the world from media-based to systems-based compliance enforcement approaches that require EMS-type strategic environmental planning by top management (see, for example, the EPA's National Environmental Performance Track program at www.epa.gov). A company with a productive EMS is positioned to take advantage of the many cost and other benefits of these new enforcement approaches.

5. *Improve prospects for long-term sustainability.* Because an EMS leads to continually improving environmental performance through P2 and other measures, companies that implement and maintain an EMS continually reduce their demands per unit of production for energy and natural resources. This improves a company's prospects for sustainability as resources become

scarcer and more expensive. It also contributes to sustainability by reducing the risk of potentially catastrophic environmental liability. Finally, it helps assure that a company's operations will not be shut down for reasons of regulatory noncompliance.

6. *Reduce operating costs.* A well-functioning EMS continually generates waste and energy minimization opportunities. As a result, a company that establishes and maintains an EMS always, in due course if not immediately, realizes substantial operating cost reductions. Such cost reductions—from an energy-saving improvement, for example—continue to accrue year after year. In many cases the accrued savings increase on an annual basis over time: increases in the costs of energy or hazardous waste disposal over the last two decades, for example, have far outpaced national inflation averages in the United States and parts of the European community.

7. *Achieve high rates of return on capital investments.* An EMS systematically identifies and prioritizes process, technology, and materials substitution investment opportunities that improve environmental performance. These opportunities are usually for P2 and energy efficiency investments, yielding returns that commonly far exceed typical rates of return on other types of investments. The financial benefits of these investments can be seen to be even greater when reduction in risk of environmental liability is factored in.

8. *Improve access to capital and lower capital and other business service costs.* Lenders tend to favor companies with an EMS because an EMS indicates better overall management and likelihood of business sustainability: they tend to lend to those companies first and at better rates. Environmental due diligence is now routine in mergers and acquisitions, and studies have shown that companies with EMSs tend to fare better in the stock market than competitors with no EMSs. Many insurance companies also now prefer commercial clients with EMSs and, depending on the nature of the insurance, may offer lower premiums than they offer to similar businesses without EMSs.

9. *Improve market access.* Governments, consumers, and major business customers (automakers are an important example) are increasingly demanding that their suppliers demonstrate a commitment to environmental responsibility. As a result, companies with EMSs have access to some markets and specific customers that is denied to competitors without EMSs.

10. *Achieve better public, community, and government relations.* A well-run EMS encourages dialogue and responsiveness to the concerns of outside parties. Some EMSs, such as the European Union's Eco-Management and Audit Scheme (EMAS), require an active stakeholder outreach program. But, even when that is not the case (ISO 14001, for example, does not require an outreach program), just having a productive EMS in place commends a company to its stakeholders as one that is environmentally responsible.

Any sound EMS will be productive; that is, it will generate impressive improvements in environmental performance, operational efficiency, and environmental risk reduction for a company. Any productive EMS will provide a company with all the benefits just listed. But, when a company's EMS has been certified to be in conformance with the requirements of the ISO 14001 international EMS standard, it tends to reap additional benefits. These additional benefits lie especially in the areas of improved capital access, improved market access, and improved public image. Additional benefits accrue in these areas because ISO 14001 certification is acquired as the result of an independent audit by an official outside EMS auditor and, therefore, is taken as proof that the company has a sound EMS in place. In other words, ISO 14001 certification gives the company's EMS visible legitimacy.

Whether or not the additional benefits of acquiring and maintaining ISO 14001 certification are worth the additional burdens this imposes continues to be a much-discussed topic. But, as of January 2002, an estimated 36,000 organizations around the world were ISO 14001 certified, and countless others had designed their own EMSs on ISO 14001 principles; so it is no wonder that many people use the terms EMS and ISO 14001 virtually as synonyms. ISO 14001 is the *only* international standard for an EMS and is useful as a basic EMS framework, whether or not a company chooses to pursue ISO 14001 certification.

The checklist in Table 4–1 is a reminder of the many drivers and benefits that a well-designed EMS with CP/P2 at its core can offer.

Table 4–1 The Benefits of an EMS with CP/P2 at Its Core

Reduced
- Raw material and energy costs
- Treatment and disposal expenses
- Associated labor costs

Many P2 strategies, such as substituting toxic materials with safer alternatives, are simple and inexpensive

Quality control techniques (e.g., ISO 9000) based on SPC methodologies have objectives identical to P2/waste minimization
- Reduced waste and variation
- Improved environmental performance
- Enhanced public image, as consumers more favorably view businesses that adopt green strategies
- Increased profits from marketing these practices

Increased productivity and efficiency
- BATs (best available technologies) can decrease raw material usage, eliminate unnecessary operations, increase throughput, reduce off-spec product generation, and improve yields

Reduced regulatory burden
- Improved environmental performance and reaching performance goals that go beyond compliance are ways to reduce regulatory burdens

Decreased liability
- Handling hazardous and toxic materials brings high liability should an accident occur
- Enterprises that substitute toxic materials with safer alternatives reduce the liability and high costs associated with an unsafe environment

Improved environmental and health quality
- Help reduce the air, water, and land pollution that results from waste generation, treatment, and disposal
- Reduce worker and resident health risks and environmental risks associated with pollutant emissions
- Conserve resources and landfill space

In summary,

- Good environmental performance makes good business sense.
- Eliminating wastes and pollution improves competitiveness and reduces liabilities.
- By focusing on all wastes and inefficiencies, precious resources are conserved and operations sustained.

A Case Study on Greening the Supply Chain

The facility (for sensitivity reasons the company will not be identified) manufactures a specialty synthetic elastomer known as EPDM (ethylene-propylene-diene-monomer) rubber. EPDM rubbers are quite versatile and fit into many specialty applications. Some examples are

- Automotive parts (e.g., trunk and window seals, weather strip, under-the-hood applications such as coolant hoses, electrical sheathing, the white beading on tires, bumpers, engine mount dampeners).
- Construction components (e.g., sealants, elastomeric roof sheeting, landfill liners).
- Health-care items (e.g., diapers, prophylactics, packaging).
- Electrical items (high-, medium-, and low-voltage electrical insulation).
- Packaging (e.g., food, pharmaceuticals).

EPDMs are specialty rubbers—they are sold into niche applications markets and command a premium because of their unique performance properties (e.g., excellent ozone and UV resistance, great weatherability, broad and low operating temperature ranges, ability to form into highly loaded compound formulations with less costly fillers like clay and carbon black).

The Process

The primary manufacturing steps are as follows:

1. Monomer and diene, along with cocatalysts are introduced into polymerization reactors. A vanadium-halide catalyst system is used and the diene of choice is ENB (ethylidene norn-bornene). Hexane serves as a hydrocarbon diluent for making a suspension of the rubber.

2. Following polymerization, the rubber suspension is quenched with water (to kill the reaction) and antioxidant and stabilizers are added (to stabilize the product).

3. The suspension then is sent to a flash drum, where it is contacted with steam and hot water, resulting in a volatilization of the hexane, which can be recovered and reused. The flashed rubber product is suspended in a hot water bath within the flash tank.

4. The recovered rubber is then sent to the "finishing stages" of the operation. Finishing involves drying and removing any residual hydrocarbons. The unit processes used for finishing include a combination of draining, application of thermal energy, and kneading/extruding, which subjects the rubber to viscous heating to aid in water vaporization.

5. From the finishing stages, the product is formed. The product forms are solid bales of rubber, crumb, and pellets.

6. The final rubber product is then packaged (either as solid bales wrapped in a protective coating, usually PE [polyethylene], or in bags for crumb and pellets).

7. The final stage in production is warehousing and filling shipping orders to worldwide customers.

The Environmental Aspects

The major environmental aspects characterizing each stage of the process are shown in Table 4–2. The weighting factor reflects the relative severity of the environmental aspect (the higher the number, the more severe the group of environmental aspects relative to each other).

Table 4-2 The Environmental Aspects of the Operation

Step	Operation	Environmental Aspect	Weighting Factor
1	Polymerization	Air emissions	6
2	Quenching	Sludge, water pollution, air emissions	12
3	Flashing and hexane recovery	Air emissions, water pollution, downtimes from flash drum plugging, solid waste, and high energy consumption	7
4	Finishing	Air emissions, water pollution, solid waste (scorched rubber), excessive energy, fires from spontaneous combustion, worker health risks from inhalation hazards	20
5	Product forming	Solid waste (off-spec product)	7
6	Packaging	Minimal solid waste, occasional worker physical injuries	4
7	Warehousing and shipping	Occasional fires, occasional worker physical injuries	8

The Root Causes and Relations to the Supply Chain of Events

On the surface, these may seem like isolated incidents, and organizations that have an EMS and P2 program in place would begin to systematically address each of these aspects. Chapter 5 summarizes a methodology for conducting CP/P2 audits, applying balances to quantify material and energy flows, and how to use the information as a basis to formulate solutions that would reduce the severity or eliminate the aspects, while achieving savings by reducing waste and inefficiencies. With this approach, the P2 team focuses on identifying and devising alternatives that eliminate pollution at the source.

While this methodological approach results in major improvements to environmental performance and provides the savings and in many cases cash flows that enable reinvestments for modernization, it has the drawbacks of limiting the benefits to within the organization and generally requires higher-cost P2 solutions than the approach presented next. While the latter provides incentives for an enterprise to seek P2 solutions and opportunities beyond the plant fence, the former provides an incentive that is more subtle. That incentive is development of closer relations between suppliers and customers, which no one should doubt improves business performance and margins.

To illustrate these points, let us consider the relationships between cause and effect for one environmental aspect, solid waste. Major solid waste is generated at several points in the process:

- During quenching, where gels and very low MW (molecular weight) by-products result.
- Solvent recovery, flash tanks have a tendency to plug, resulting in costly downtimes and discarding the product.
- During finishing, too severe drying conditions can scorch (burn) the rubber, making an off-spec product.
- Sales, customers reject product as being off-spec.

In a conventional approach, a P2 team would focus on reducing sources of waste and inefficiency by focusing on opportunities at the source. As examples, quenching operations can be improved through the use of in-line mixers (a low-cost P2 solution) to enhance mixing, which would terminate reactions faster and reduce the possibility of making a gelled product. In the flash drum (FD), the installation of baffles and increasing impeller size would enhance mixing, thereby eliminating rubber agglomeration and plugging. These would be moderate-cost P2 solutions. In finishing, installing low-shear extruders and kneaders would eliminate scorching. This would require a high-level P2 investment. All these actions would help improve product quality and thereby making fewer off-spec products and reducing customer rejects. They most certainly would improve the rubber supplier's productivity and reduce the need for landfill disposal.

All these are good engineering alternatives with overall attractive returns on investment (ROIs). The low-cost/no-cost P2 alternatives have almost immediate paybacks, and the high-cost investment would require less than five years payback for a $6.5 million investment. The high-cost investment is even more attractive when one considers that frequent fires occur for certain grades (product variations) that easily scorch.

But, other alternatives bring even greater benefits. By understanding what the customer and suppliers are doing, external cause and effects are identified, which have an impact not only to the plant, but other players along the supply chain.

Engineers involved in the exercise quickly realized this when they began working closely with their customers. Customers insisted that ultranarrow MWD (molecular weight distribution) rubbers cure faster, which translates into increased throughput during compounding and parts forming operations. However, the exact opposite was found for a number of key rubber grades. While it is true that narrow MWD rubbers cure faster, these rubbers are more difficult to compound and process, so much so that for some production campaigns as much as 15% of a rubber compound batch for manufacturing car weather strip had to be rejected, creating a serious solid waste problem. By working with customers to define an optimum cure-processability window of operation, a slightly broader MWD product was made. This totally eliminated the need for any hardware changes within the plant, eliminated fires altogether, reduced the energy requirements in finishing operations by more than 30%, and reduced a solid waste problem in the customer's shop by 98%. In total, the net savings to customer and plant amounted to nearly $700,000 per year.

But what about the upstream side of the supply chain? By working closely with catalyst and termonomer suppliers, the company was able to identify and isolate poisons that cause side reactions from occurring during polymerization. By eliminating or minimizing the presence of catalyst poisons, fewer side reactions occur and hence both less off-spec production and shorter turnaround times for reactor clean-out operations are required. This resulted in less time required for grade switching, meaning polymerization was much more efficient (by as much as 25%). The higher productivity reduced air emissions by 10%, downstream solid wastes by 15%, and wastewaters by more than 25%. But the upstream suppliers also received benefits and improved their environmental performance. As one example, the catalyst supplier was able to remove cyanide from its product, eliminating altogether liabilities and a serious pollutant. In addition, the supplier itself benefited from the effort since it could supply a higher purity product to its customers.

Greening the Supply Chain

Why should a company invest in greening the supply chain? According to General Motors (GM) "Working together with our suppliers, we can accomplish much more to improve the environment than GM can alone."[1]

Greening the supply chain refers to buyer companies requiring a certain level of environmental responsibility in core business practices of their suppliers and vendors. Many businesses have internal standards, policies, and environmental management systems that govern their own environmental performance and efficiency. If suppliers do not abide by the same standards, the buyer company may be buying and using products that do not meet their own standards. This can lead to liability.

A supply chain can be complex, with environmental issues occurring at the second- and third-tier supplier levels. Some companies may also attempt to work the opposite spectrum of the supply chain, by educating their customers on the environmental benefits of their products.

According to an enlightening study conducted by Business for Social Responsibility (BSR),[2] many companies have received requests from their corporate customers to address environmental issues. The study also discusses benefits and challenges from the

---

[1] See http://www.pprc.org/pubs/grnchain/intro.cfm.
[2] See http://www.getf.org/file/toolmanager/O16F15429.pdf.

suppliers' perspective. Several suppliers noted that their efforts toward meeting the customers' requirements have resulted in advantages such as cost reductions, greater operational efficiency, enhanced value to customers, increased sales, positive media attention, and positive ratings from socially responsible investment groups.

When customers and their supply chains collaborate on environmental and efficiency improvements, the significant benefits of partnering are derived. Some of these are

- Suppliers know the product better than the buyers and can maximize efficiencies and minimize associated wastes.
- Multiple perspectives (or different expertise areas) are better than one when it comes to designing greener products and processes.
- Teamwork strengthens the customer-supplier relationship.
- Shared savings and mutual benefits make such efforts worthwhile.

Simple examples of partnering between supplier and customer include collaborating on green design and manufacturing projects; sharing tools used for environmental improvement; researching alternative materials, products, equipment, and processes that have lower life-cycle impacts; managing of inventories by the supplier (e.g., chemicals, cleaning supplies, lab supplies, office supplies); and devising ways to take back and recycle or refurbish end-of-life items and packaging.

Buyers can use their purchasing power to influence suppliers and help create a more reliable market for greener products. By specifying purchasing preferences for products that have lesser environmental impact, companies can spur manufacturers to rely on cleaner production technologies, practices and materials, design for the environment (DfE), consume less energy or water, minimize waste, or create less toxic products or emissions. All these are factors that affect on both sustainability and competitiveness.

Examples of major corporations that focus on greening efforts along the supply chain are

- Bristol-Myers Squibb, a developer of beauty, nutritional, and pharmaceutical products, works with major suppliers to align their environmental management systems with the requirements of ISO 14001.
- General Motors and other auto manufacturers (e.g., Ford and Toyota) require suppliers to certify implementation of an environmental management system or ISO 14001 compliance.
- Hewlett-Packard worked with other electronics manufacturers on a standardized questionnaire for environmental evaluation of supplier practices, in attempts to minimize duplication of suppliers' efforts for the multiple customers conducting similar evaluations.
- Johnson Controls responded to Ford's ISO 14001 mandate and in turn began requiring that its suppliers become certified.
- Nokia published environmental guidelines for suppliers.

Greening the supply chain is best known as *supply chain management*. This refers to the oversight of materials, information, and financial flows inside and outside an organization. Flows move from supplier to manufacturer to wholesaler to retailer to consumer. Supply chain management is concerned with planning, implementing, and controlling the flows of raw materials, inventory, and finished goods from the point of origin to the point of consumption and covers purchasing, the manufacturing process, and customer delivery. The concept is also referred to as *environmental supply chain management* or *responsible supply chain management*. Green supply chain management is designed to incorporate environmental considerations into decision making at each stage of material management including postconsumer disposal. Specific terminology used by experts or emerging experts in this field include:

Design for the environment. A design for the environment approach helps industry eliminate or minimize the environmental impacts of products during the design stage. It can involve reducing the toxicity of a product, extending the life of a product, extending the life of the material used, improving the selection of materials, and reducing the energy and material intensity required to produce, use, and dispose of the product (*Pollution Prevention Planning Handbook*, Environment Canada).

- *Eco-efficiency.* Introduced earlier, the term refers to a management strategy that links financial and environmental performance to create more value with less ecological impact. Eco-efficiency is the ability to meet economic performance goals while using environmentally benign modes of doing business (World Business Council for Sustainable Development).
- *Environmental preferable purchasing.* Environmental preferable purchasing (EPP) refers to choosing products or services that have a reduced environmental impact. Considering products that are energy efficient, limit the use of toxic elements, and reduce waste are some of the criteria for EPP (U.S. EPA). Corporate and institutional consumers can incorporate environmental requirements into their product and packaging specifications (*Pollution Prevention Planning Handbook*, Environment Canada).
- *Extending producer responsibility.* Extending producer responsibility (EPR) is a policy option that extends the responsibility of producers to include the environmental impacts of their products over the entire product life cycle. This usually occurs once a product has been discarded by the consumer. EPR strategies include product take back, recycling, and disposal (INFORM Inc.).
- *Life-cycle perspective.* A life-cycle perspective prevents the environmental impacts of a product or activity across its entire life by planning products and activities from inception, final use, and disposal—"cradle to grave" (P2WIN). Tools such as life-cycle management (LCM) and life-cycle assessment (LCA) can help evaluate options to ensure that material substitution or process changes do not shift environmental and financial impacts to another stage along the life cycle (*Pollution Prevention Planning Handbook*, Environment Canada).
- *Materials management.* The business process that supports the complete cycle of material flows, including internal control of

purchasing, production materials, planning, warehousing, and shipping and distributing finished products (*The Lean and Green Supply Chain*, U.S.EPA, January 2000).

Product take back. This includes activities to obtain used products from industrial customers or consumers, then recycle or reuse these products (*The Lean and Green Supply Chain*, U.S. EPA, January 2000).

Resource efficiency. Resource efficiency is a measure of how effectively resources such as energy, water, and materials are being used to meet needs. It improves as fewer resources are used and enhanced and more diverse services and products are provided over longer periods of time (www.ConservationEconomy.net). Improvements in resource efficiency can help break the link between economic growth and pollution emissions (Organization for Economic Cooperation and Development).

The first step in greening a supply chain brings us back to the first step on an EMS: an initial environmental review. Most companies have little knowledge of the effects that their activities, products, and services have on the environment and of the environment on them, especially along the supply chain. All types of business and industry are subject to increasing pressures to demonstrate their commitment to improved environmental performance. These pressures are imposed by increasing environmental legislation, the financial markets, and customers through the supply chain. Companies that actively identify and systematically manage environmental impacts and risks are better placed to ensure compliance with environmental regulations, to reduce environmental liability, to meet the demands of the supply chain, and to take advantage of environmental market opportunities. The IER is the first step toward establishing a long-term strategy for a company to manage environmental impacts.

Since an important driver for many companies to consider cooperative programs along the supply chain is to reduce the risks associated with environmental management, an additional term that may help to place further importance on the IER for suppliers,

Improving Competitiveness through an EMS 145

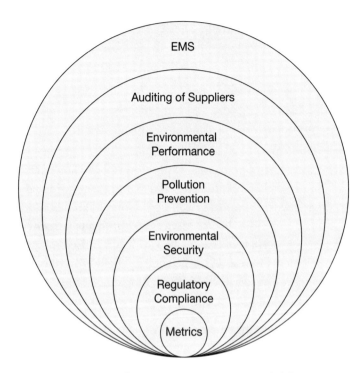

Figure 4–1 Elements of supply chain environmental risk management.

distributors, and even customers is *supply chain environmental risk management*, or SCERM. The interactions along the supply chain are illustrated in Figure 4–1.

In examining Figure 4–1, the elements listed are identical to the ones considered when we develop the EMS within our company. We are simply extending the EMS to our suppliers, distributors, and customers in some cases. Clearly, the application of an EMS is more far-reaching than within our own universe of manufacturing. The benefits of an EMS on a local basis are extremely significant. The skeptic can review Chapters 6 and 7 as well as simply surf the Web for almost countless examples. The extension of an EMS along the supply chain unquestionably has global implications in terms of sustainability, market robustness, and growth.

5

Conducting Cleaner Production Audits

Introduction

As the reader has grasped by now, an EMS regiments an organization into managing environmental aspects within the framework of a set of standardized procedures and practices. Those readers who have been through an ISO 9001 program recognize that a key factor in the ability of an enterprise to reduce defects from production is by systematically reducing variation in the way a product is manufactured. Operator procedures, process operating conditions, factors influencing the management of materials and energy, and all factors that affect product variation are always being improved on a continual basis, striving to reduce variation and control manufacturing campaigns within acceptable limits.

An EMS, and especially ISO 14001, is designed to function in an identical manner. Under an EMS that focuses on pollution prevention, *zero waste* is as much a goal as *zero defects* in a quality control program. And just as the goal of zero defects is pursued by a systematic methodology built around auditing principles, so is the goal of zero waste.

While there are a number of ways to approach the audit, it is best to apply a standardized methodology that results in procedures that can be used applied routinely and adopted easily in different parts of a facility and other facilities owned and operated by a company. Conducting a pollution-prevention-type audit involves both acquiring and transferring skills effectively throughout an organization. I have found that the best approach to accomplishing this is to limit the initial audit to one aspect or production platform within a facility. Once procedures, practices, and skills have been successfully acquired and positive results achieved, the skills can be transferred to other divisions, enabling independent functioning but communicating CP/P2 teams or cells to function on their own.

This chapter provides a rigorous 20-step procedure for conducting in-plant CP/P2 assessments. The chapter is organized and written such that the audit team leader can instruct and train members on a generalized approach to performing the audit and analysis of information. The last section of this chapter contains a toolkit comprised of worksheets that help facilitate the P2 on-site assessment. The worksheets are contained on the CD Toolkit for printing or data entry. The CD version of the Toolkit is described at the end of this chapter.

In-Plant Assessment and Pollution Prevention

In-plant environmental assessments should initially focus on low-cost/no-cost corrective actions aimed at improved environmental management with accompanying economic benefits to the enterprise. These generally will result in the identification and capture of low-hanging fruit that give immediate, although in most cases modest, returns. While skeptics may argue that most of the housekeeping and obvious P2 opportunities that fall in this category have been addressed by U.S. firms because of the heightened level of environmental awareness and aggressive enforcement, I found the contrary to be the case, particularly for SMEs (small and medium-size enterprises).

When conducting such initial assessments, the reasons or objectives for doing so should be understood by the employees. It is important to convey the following messages:

- Pollution prevention (CP/P2) focuses attention away from the treatment and disposal of wastes and toward the elimination or reduction of undesired by-products within the production process itself.
- In the long run, pollution prevention through waste minimization and cleaner production is more cost-effective and environmentally sound than traditional pollution control methods.
- Pollution prevention techniques apply to all manufacturing processes and range from relatively easy operational changes and good housekeeping practices to more extensive changes, such as the substitution of toxic substances, the implementation of clean technology, and the installation of state-of-the-art recovery equipment.
- Pollution prevention can improve plant efficiency, enhance the quality and quantity of natural resources for production, and make it possible to invest more financial resources in economic development.

All outputs from a manufacturing facility can be put into two classifications, *product* and *waste*: anything that the customer pays for is product; everything else that leaves the facility is waste. Ideally, manufacturing activities should produce zero waste. In reality, industry must strive to reduce the waste from manufacturing, since this represents an inefficient use of resources.

We can argue that all waste can be indirectly associated with pollution, since the management of waste consumes resources that would not otherwise be used, and pollution is often generated in these waste management activities.

Prevention is the act of taking advance measures against something possible or probable. Prevention is generally contrasted with control or cure.

- Design for *quality prevents* defects, while *inspection controls* defects.

- In general, the effort, time, and money associated with prevention is less than that with control or cure. This idea is captured in the maxim: an ounce of prevention is worth a pound of cure. In many cases, it is worthwhile for industry to prevent pollution rather than control it.

Pollution prevention is any practice that

- Reduces the amount of any hazardous substance, pollutant, or contaminant reentering any waste stream or otherwise released into the environment prior to recycling, treatment, and disposal.

- Reduces the hazards to public health and the environment associated with the release of such substances, pollutants, or contaminants.

- Reduces or eliminates the creation of pollutants through (1) increased efficiency in the use of raw materials or (2) protection of natural resources by conservation.

As another way of saying this, pollution prevention is any action that reduces or eliminates the creation of pollutants or wastes at the source, achieved through activities to promote, encourage, or require *changes in the basic behavioral patterns of industry*. And, as stated earlier, in common terminology, the moves toward this goal are

Cleaner Production
∇
Clean Technology
∇
Waste Reduction
∇
Waste Prevention
∇
Eco-efficiency
∇
Waste Minimization

Let's consider the following example of CP/P2.

Pollution Prevention in Auto Painting

Problem
It is common for auto companies to change paint color with each car that goes through the paint process. As a result, old paint must be purged from the lines before painting each car. This results in excess paint sludge waste and fugitive emissions of toluene and xylene. Additionally, the purging and refilling qualifies as a setup activity that adds time to the process.

The reader may recall the synthetic rubber production example in Chapter 3, in which grade production sequencing was used to reduce the turnaround times between reactor cleanout batches. The same general methodology in the form of a simple operational change has major waste reduction and savings benefits across two entirely different industry sectors.

Pollution Prevention Solution 1
Block painting, the process of painting batches of like-colored cars, is a manufacturing process change that reduces the purged paint sludge and solvent emissions. Block painting decreases not only the waste but also the setup time involved in the process.

Pollution Prevention Solution 2
Cars can be painted without toxic toluene and xylene solvents. Electrostatic painting can adhere paint to treated metal. While the scrubber represents treatment and block painting represents waste reduction, shifting to the electrostatic painting process represents *pollution prevention by design*.

The Environmental Management Hierarchy

Environmental management involves several strategies for dealing with wastes. It is important to convey the relative importance of the hierarchy of strategies to employees. *Strategies that reduce or eliminate wastes before they are created are preferable to those that deal with treating or disposing wastes that are already generated.* By way of review, the hierarchy is

Prevention. The best waste reduction strategy is one that keeps waste from being formed in the first place. Waste prevention may in some cases require significant changes to a process, but it provides the greatest environmental and economic rewards.

Recycling. If waste generation is unavoidable in a process, then strategies that minimize the waste to the greatest extent possible should be pursued, such as recycling and reuse.

Treatment. When wastes cannot be prevented or minimized through reuse or recycling, strategies to reduce their volume or toxicity through treatment can be pursued. While end-of-

Table 5–1 Sample Applications Illustrating the Environmental Management Hierarchy

Priority	Method	Examples	Applications
1	Prevention (source reduction)	• Process changes • Design of products that minimize environmental impacts • Source elimination	• Modify process to avoid/reduce solvent use • Modify product to extend coating life
2	Recycling	• Reuse • Reclamation	• Solvent recycling • Metal recovery from spent bath • Volatile organic recovery
3	Treatment	• Stabilization • Neutralization • Precipitation • Evaporation • Incineration • Scrubbing	• Thermal destruction of organic solvents • Precipitation of heavy metals from spent plating bath
4	Disposal	• Disposal at a facility	• Land disposal

pipe strategies can sometimes reduce the amount of waste, they are not as effective as preventing the waste in the first place.

Disposal. The last strategy to consider is alternative disposal methods. Proper waste disposal is an essential component of an overall environmental management program; however, it is the least effective technique.

This hierarchy is summarized in the example applications matrix presented in Table 5–1.

Reasons to Prevent Pollution

Most countries need to balance economic growth with environmental protection. It is universally recognized that economic development and the health and welfare of a society are closely linked to proper management of a country's natural resources and the environment. In these countries, pollution prevention offers the government and industry a way to manage the impacts of industrial growth on the environment while enabling economic development.

Pollution prevention addresses three important components of the environmental protection/economic development issue:

Prevention offers a better solution for environmental management than end-of pipe pollution solutions.

Quality encourages evaluation of production processes and product quality.

Cost improves a facility's bottom line by reducing treatment costs, saving on material and resource inputs, and reducing risk and liability insurance.

Dealing with environmental wastes through end-of-pipe measures (such as wastewater treatment systems, hazardous waste incinerators and other treatment technologies, secure landfills, monitoring equipment, solid waste hauling equipment, air pollution control equipment, and catalytic converters) has proven very costly and does not address all environmental problems. CP/P2 offers industry the advantages of

1. Less need for costly pollution control equipment.
2. "Getting ahead" of environmental regulations.
3. Reduced reporting and permitting requirements.
4. Less operation and maintenance of pollution control equipment.

Improving Product Quality through Total Quality Environmental Management

The process of identifying pollution prevention opportunities also provides the opportunity to identify measures to improve product quality. The in-plant environmental (pollution prevention) assessment requires the enterprise to examine its production process in-depth. Finding ways to reduce wastes also requires a study of the root causes for generating wastes and finding improvements to processes.

Total quality management (TQM) is the management system developed to achieve the goal of high product and service quality. The management elements of TQM include

1. Customer focus.
2. Continuous improvement.
3. Teamwork.
4. Strong management commitment.

At first glance, TQM seems unrelated to these environmental concerns. Yet the inherent strengths of the TQM methodology can effectively address many environmental issues. Professionals who apply TQM concepts to environmental issues have coined the term *total quality environmental management* (TQEM). TQEM is a logical method for achieving pollution prevention.

In the context of quality, the customer is defined as the person or entity who employs the "product and service characteristics." Customers fall into two categories, internal and external. The internal customer is the next person in the production chain, while the external customer is the end-user of the product.

If the definition of a *customer* is expanded to include those people and environments affected by the production process waste, total quality management requires us to understand the impact of this waste on those customers and take steps to reduce it.

Consider the following case study.

Case Study: Trichloroethylene

Recognizing the Customer

Many industries use the solvent trichloroethylene (TCE) in their operations. This toxic chemical must be contained in a closed system, as releases of TCE can be fatal. Such releases often require the evacuation of the facility. Plant workers are the unwilling internal customers of TCE fumes. The external environment is also an unwilling customer. Groundwater and surface water bodies can be effected by effluents. Aquatic life in the surface water body and people dependent on the river for drinking water or recreational purposes are unwilling customers as well as those dependent on groundwater sources.

Continuous Improvement

Quality can be built into, not inspected into, the product. This requires the producer to continuously identify and eliminate the root cause of the impediments to quality. Continuous improvement is also the key to reducing the environmental impacts of the production process.

The traditional approach to industrial waste has been to view it as a necessary by-product of manufacturing. While production generates the waste, the responsibility to dispose of the waste in a safe and legal manner usually falls on the environmental engineering department. Because environmental engineers receive the waste after it has been created, they are not intimately familiar with the processes that create it. Further, because waste reduction is not a component of their performance review, environmental engineers have no institutional motivation to reduce the waste.

TQEM is the logical method for producing the results of pollution prevention. Pollution prevention calls for industry to prevent pollution wherever possible. By employing a customer focus and classifying the waste itself and the activities required to control it as non-value-added, TQEM calls for waste generation to be brought to a minimum. Operators and process engineers, not environmental engineers, are responsible for identifying and eliminating the root causes of process waste. Employing the continuous improvement approach, zero waste is as much a goal as zero defects.

As a result of TQEM projects, product quality often improves while waste is reduced. TQEM efforts empower employees to become more familiar with all aspects of the process and not just those associated with production. When forced to question wastes from the process, improvements to quality characteristics can result.

Teamwork

The team approach allows all factors of the environmental issue to be considered. Accountants are familiar with cost considerations, product engineers are familiar with quality considerations, process and chemical engineers are familiar with feasibility considerations, and environmental engineers are familiar with environmental impacts. Because environmental engineers are trained to deal with waste after it has been generated, and not in methods of preventing it from being created in the first place, engineers with knowledge of the process characteristics must be involved.

Ford Company Team Eliminating TCE through Continuous Improvement

Degreasing certain aluminum components with TCE has required extensive safety mechanisms and procedures. Building better containment systems reduces the risk of exposure but does not get to the root cause of the problem—the use of TCE.

With this in mind, the Ford team looked for a TCE-free solution to degreasing radiator coils. They formed a team that included a chemical engineer, an environmental engineer, a process engineer, an accountant, and a product engineer. The variety of backgrounds on the team ensured that all the pertinent issues of cost, product quality, process feasibility, and environmental impact were addressed. The team designed an aqueous degreasing system (i.e., soap and water) to replace the TCE. Not only is the toxic chemical removed from the plant, but the water in the new system is recycled as well. Significantly, the aqueous degreaser exhibits better quality characteristics than the TCE degreaser.

This project is an example of improved quality, reduced cost, and reduced environmental impact. Certainly not all projects prove so fruitful. Some clean alternatives may cost more than their polluting rivals, but that cost must be balanced with the benefits of the environmental improvement. To justify this viewpoint, one needs only to look to the increasing expectations of external customers for "environmentally friendly" products.

Strong Management Commitment

The three of the elements of TQM—customer focus, continuous improvement, team approach—readily apply to environmental issues. As in traditional TQM settings, the last—strong management commitment—is perhaps the most important.

No TQEM program will succeed without the commitment of senior management. Senior management, those who built their careers when waste was seen as a necessary by-product, must come to understand that both internal and external customer expectations include environmentally conscious products and processes. They must learn to see the value of applying TQEM to get to the root causes of waste and call on the cross-disciplinary teams to employ continuous improvement to implement ever "cleaner" solutions.

Improving Profitability

In many cases, pollution prevention measures can have clear environmental benefits in terms of pollution that is not generated, reduction in the toxic materials used in the production process, and savings in energy use and other raw materials. Savings can accrue in five areas:

1. A company can save on raw materials.
2. A company can save on labor costs.
3. Disposal costs can be reduced or eliminated.
4. A facility can save on waste handling and treatment costs both in its own use of labor to collect, store, and process wastes and incurred costs to transport wastes off-site.
5. Decreasing the amount of toxic materials used, handled, and transported at a facility can reduce its future liability costs.

Steps to Conducting CP/P2 Audits

This section describes a step-by-step approach for carrying out the in-plant CP/P2 assessment following pollution prevention principles. It is designed to be generic, to apply to a broad spectrum of industry. The approach comprises three phases:

1. A preassessment phase for assessment preparation.
2. A data collection phase to derive a material balance.
3. A synthesis phase where the findings from the material balance are translated into a waste reduction action plan.

It is possible that not all the assessment steps will be relevant to every situation. Similarly, in some situations, additional steps may be required.

Phase 1. The Preassessment

Step 1. Assessment Focus and Preparation

A thorough preparation for a pollution prevention assessment is a prerequisite for an efficient and cost-effective study. Of particular importance is to gain support for the assessment from top-level management and for the implementation of results; otherwise, no real action will occur.

The pollution prevention assessment team should be identified. The number of people required on an assessment team depends on the size and complexity of the processes to be investigated. A pollution prevention assessment of a small factory may be undertaken by one person with contributions from the employees. A more-complicated process may require at least three or four people: technical staff, production employees, and an environmental specialist. Involving personnel from each stage of the manufacturing operations increases employee awareness of waste reduction and promotes input and support for the program.

A pollution prevention assessment may require external resources, such as laboratory and possibly equipment for sampling and flow measurement. You should attempt to identify external resource requirements at the outset of the project.

Analytical services and equipment may not be available to a small factory. If this is the case, investigate the possibility of forming pollution prevention associations with other factories or industries; under this umbrella, the external resource costs can be shared.

It is important to select the focus of your assessment at the preparation stage. You may wish the pollution prevention assessment to cover a complete process or concentrate on a selection of unit operations within a process. The focus will depend on the objectives of the pollution prevention assessment. You may wish to look at waste minimization as a whole or concentrate on particular wastes; for example,

- Raw material losses.
- Wastes that cause processing problems.
- Wastes considered hazardous or for which regulations exist.
- Wastes for which disposal costs are high.

A good starting point for designing a pollution prevention assessment is to determine the major problems or wastes associated with your particular process or industrial sector. All existing documentation and information regarding the process, the plant, or the regional industrial sector should be collated and reviewed as a preliminary step. Regional or plant surveys may have been undertaken; these could yield useful information indicating the areas for concern, and they will also show gaps where no data are available. The following prompts give some guidelines on useful documentation.

- Is a site plan available?
- Are any process flow diagrams available?
- Have the process wastes ever been monitored—do you have access to the records?
- Do you have a map of the surrounding area, indicating watercourses, hydrology, and human settlements?
- Are any other factories or plants in the area that may have similar processes?

The following list has other general data that may be collated quickly and be useful as orientation material.

- What obvious wastes are associated with your process?
- Where is water used in greatest volume?
- Do you use chemicals that have special instructions for their use and handling?
- Do you have waste treatment and disposal costs and what are they?
- Where are your discharge points for liquid, solid, and gaseous emissions?

The plant employees should be informed that the assessment will be taking place, and they should be encouraged to take part. The support of the staff is imperative for this type of interactive study. It is important to undertake the assessment during normal working hours, so that the employees and operators can be consulted, the equipment can be observed in operation, and most important, wastes can be identified and quantified.

Step 2. Listing Unit Operations
Your process will comprise a number of unit operations. A *unit operation* may be defined as an area of the process or a piece of equipment

Conducting Cleaner Production Audits

where materials enter production, a function occurs, and materials leave that portion of production, possibly in a different form, state, or composition. For example, a process may comprise the following unit operations: raw material storage, surface treatment, rinsing, painting, drying, product storage, and waste treatment.

Any initial site survey should include a walk around the entire manufacturing plant to gain a sound understanding of all the processing operations and their interrelationships. This will help the assessment team decide how to describe a process in terms of unit operations. During this initial overview, it is useful to record visual observations and discussions and make sketches of process layout, drainage systems, vents, plumbing, and other material transfer areas. These help ensure that important factors are not overlooked.

The assessment team should consult the production staff regarding normal operating conditions. The production or plant staff are likely to know about waste discharge points, unplanned waste generating operations such as spills and washouts, and give the assessors a good indication of actual operating procedures. Investigations may reveal that night-shift procedures are different from day-shift procedures; also, a plant may disclose that actual material handling practices are different from those set out in written procedures.

A long-standing employee could provide some insight into recurring process problems. In the absence of any historical monitoring, this information can be very useful. Such employee participation however must be a blame-free process; otherwise, it will not be as useful as it could be.

During the initial survey, note imminent problems that need to be addressed before the assessment is complete.

The assessment team needs to understand the function and process variables associated with each unit operation. Similarly, all the available information on the unit operations and the process in general should be collated, possibly in separate files. It is useful to tabulate this information, as shown in Table 5–2.

Identification of materials handling operations (manual, automatic, bulk, drums, etc.) covering raw materials, transfer practices,

Table 5-2 Example of Identifying Process Variables That Affect Waste Generation

Unit Operation	Function	File Number
(A) Surface Treatment	Surface treatment of glass bottles: 100 m^3 spray chamber, 6 jets, 100 gal/min pumps	1
(B) Rinsing	Washing glass bottles before labeling	2

and products is also an important aspect that could usefully be included in this tabulation as a prelude to development of a materials balance (phase 2).

Step 3. Constructing Process Flow Diagrams

By connecting the individual unit operations in the form of a block diagram you can prepare a process flow diagram. Intermittent operations, such as cleaning, makeup, or tank dumping may be distinguished by using broken lines to link the boxes.

- For complex processes, prepare a general flow diagram illustrating the main process areas and, on separate sheets of paper, detailed flow diagrams for each main processing area.
- Now, decide on the level of detail that you require to achieve your objectives.
- It is important to realize that the less detailed or larger scale the assessment becomes, the more information is likely to be lost or masked by oversimplification. Establishing the correct level of detail and focusing in on specific areas is very important at an early stage.
- Pay particular attention to correcting any obvious waste that can be reduced or prevented easily, before proceeding to the development of a material balance (phase 2). By making simple changes at this early stage, the resultant benefits will help enlist the participation and stimulate the enthusiasm of employees for the total pollution prevention assessment and reduction program.

Phase 1. Summary

- At the end of the pollution prevention assessment's preassessment stage, the assessment team should be organized and aware of the objectives of the pollution prevention assessment.
- Plant personnel should have been informed of the assessment purpose to maximize cooperation among all parties concerned.
- Any required financial resources should have been secured and external facilities checked out for availability and capability.
- The team should be aware of the overall history and local surroundings of the plant.
- The scope and focus of the pollution prevention assessment should have been established, and a rough timetable worked out to fit in with production patterns.
- The assessment team should be familiar with the layout of the processes within the plant and should have listed the unit operations associated with each process. Sources of wastes and their causes also should have been identified.
- It should be possible to draw process flow diagrams highlighting those areas to be covered in the pollution prevention assessment.
- Any very obvious waste-saving measures that can be introduced easily should be implemented immediately.
- The findings of the phase 1 investigations could usefully be presented to the management in the form of a brief preassessment report to reaffirm their commitment into the next phase.

Phase 2. Material Balance: Process Inputs and Outputs

A *material balance* is defined as a precise account of the inputs and outputs of an operation.

This phase describes a procedure for the collection and arrangement of input and output data. The procedure can be applied to derive the material balance of a plant, a process, or a unit operation. Figure 5–1 is an example set of components that need to be quantified to derive a material balance. Note that infrequent outputs (e.g.,

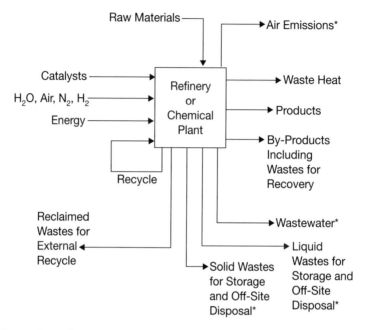

Figure 5–1 Conceptualized material balance (* notes the long-term liabilities for the company).

the occasional dumping of an electroplating bath) may be as significant as continuous daily discharges.

Step 4. Determining Inputs

- Inputs to a process or a unit operation may include raw materials, chemicals, water, air, and power. The inputs to the process and to each unit operation need to be quantified.

- As a first step toward quantifying raw material usage, examine purchasing records; this rapidly gives us an idea of the sort of quantities involved.

- In many situations the unit operations where raw material losses are greatest are raw material storage and transfer. Look at these operations in conjunction with the purchasing records to determine the actual net input to the process.

- Make notes regarding raw material storage and handling practices. Consider evaporation losses, spillages, leaks from underground storage tanks, vapor losses through storage tank pressure-relief vents and contamination of raw materials. Often these can be rectified very simply.
- Record raw material purchases and storage and handling losses in a table in order to derive the net input to the process (See Table 5–3 as an example below).
- Once the net input of raw materials to the process has been determined, we can proceed with quantifying the raw material input to each unit operation.
- If accurate information about raw material consumption rates for individual unit operations is not available, then we need to take measurements to determine average figures.
- Measurements should be taken for an appropriate length of time. For example, if a batch takes one week to run, then

Table 5–3 Example: Raw Material Storage and Handling Losses

Raw Material	Estimated Raw Material Losses (annual)	Raw Material Purchased (annual)	In-House Inventory Maintained	Average Length of Inventory Storage (annual)	Negative Impacts
1. Activated carbon for product color control	100 kg	5,500 kg	500 kg	1 month	Inventory inadequate, additional purchases needed or identify root cause for losses and eliminate
2.					
3.					

measurements should be taken over a period of at least three weeks; these figures can be extrapolated for monthly or annual figures.

Note: Some quantification is possible by observation and some simple accounting procedures.

For solid raw materials, ask the warehouse operator how many sacks are stored at the beginning of the week or prior to unit operation, then ask again at the end of the week or unit operation. Weigh a selection of sacks to check compliance with specifications.

For liquid raw materials, such as water or solvents, check storage tank capacities and ask operators when a tank was last filled. Tank volumes can be estimated from the tank diameter and tank depth. Monitor the tank levels and the number of tankers arriving on-site.

While investigating the inputs, talking to the staff, and observing the unit operations in action, the pollution prevention assessment team should be thinking about how to improve the efficiency of unit operations.

Consider the following questions:

- Is the size of the raw material inventory appropriate to ensure minimization of material-handling losses?
- Could the transfer distances between storage and process or between unit operations be reduced to minimize potential wastage?
- Do the same tanks store different raw materials depending on the batch product? Is there a risk of cross-contamination?
- Are sacks of materials emptied or is some material wasted?
- Are viscous raw material used on-site; is it possible to reduce residual wastage in drums?
- Is the raw material storage area secure? Could a building be locked at night, or could an area be fenced off to restrict access?
- How could the raw materials be protected from direct sunlight or heavy downpours?
- Is dust from stockpiles a problem?

- Is the equipment used to pump or transfer materials working efficiently? Is it maintained regularly?
- Could spillages be avoided?
- Is the process adequately manned?
- How could the input of raw materials be monitored?
- Are any equipment items in obvious need of repair?
- Are pipelines self-draining?
- Is vacuum pump water recirculated?

The energy input to a unit operation should be considered at this stage; however, energy use deserves a full assessment in its own right. For pollution prevention assessment purposes make note of the energy source and whether waste reduction could reduce energy costs. If energy usage is a particularly prominent factor, maybe you should recommend that an energy assessment be undertaken. Input data should be recorded on your process flow diagram or in tabular form as shown in Table 5–4.

Water is frequently used in the production process, for cooling, gas scrubbing, washouts, product rinsing, and steam cleaning. This water usage needs to be quantified as an input.

Table 5–4 Example of Energy Usage Tabulation

Unit Operation	Raw Material 1 (mm^3/annum)	Raw Material 2 (tons/annum)	Water (mm^3/annum)	Energy Source
Surface treatment (A)				
Rinse (B)				
Painting (C)				
Total raw material used in all unit operations				

Some unit operations may receive recycled wastes from other unit operations. These also represent an input.

Steps 5 and 6 describe how these two factors should be included in the in-plant environmental (pollution prevention) assessment.

Step 5. Recording Water Usage

The use of water, other than for a process reaction, is a factor that should be covered in all pollution prevention assessments. The use of water to wash, rinse, and cool often is overlooked, although it represents an area where waste reductions frequently can be achieved simply and cheaply.

Consider these general points about the site water supply before assessing the water usage for individual units.

- Identify water sources.
- Is water absorbed directly from a borehole, river, or reservoir; is water stored on-site in tanks or a lagoon?
- What is the storage capacity for water on site?
- How is water transferred—by pump, by gravity, manually?
- Is rainfall a significant factor on-site?

For each unit operation, consider the following:

- What is water used for in each operation—cooling, gas scrubbing, washing, product rinsing, dampening stockpiles, general maintenance, safety quench, etc.?
- How often does each action place?
- How much water is used for each action?

It is unlikely that the answers to these questions will be readily available—you may need to undertake a monitoring program to assess the use of water in each unit operation. Again, the measurements must cover a sufficient period of time to ensure that all actions are monitored. Pay particular attention to intermittent actions, such as steam cleaning and tank washout: water use is often indiscriminate during these operations. Find out when these actions are undertaken so that detailed measurements can be made.

Table 5–5 Example of Water Usage Tabulation

	Cleaning	Steam	Cooling	Other
Unit Operation A				
Unit Operation B				
Unit Operation C				

Note: All measurements are in standard units, for example m^3/annum or m^3/day.

Record water usage information in a tabular form (as in Table 5–5) and ensure that the units used to describe intermittent actions indicate a time.

Using less water can be a cost-saving exercise. Consider the following points while investigating water use:

- Tighter control of water use can reduce the volume of wastewater requiring treatment and result in cost savings. In the extreme, it can sometimes reduce volumes and increase concentrations to the point of providing economic material recovery in place of costly wastewater treatment.
- Attention to good housekeeping practices often reduces water usage and, in turn, the amount of wastewater passing to drains.
- The cost of storing wastewater for subsequent reuse may be far less than the treatment and disposal costs.
- Countercurrent rinsing and rinse-water reuse are useful tips for reducing water usage.

Step 6. Measuring Current Levels of Waste Reuse and Recycling

Some wastes lend themselves to direct reuse in production and may be transferred from one unit to another; others require some modifications before they are suitable for reuse in a process. These reused waste streams should be quantified.

If reused wastes are not properly documented, they may be double-counted in the material balance, particularly at the process or complete plant level; that is, a waste might be quantified as an output from one process and an input to another.

The reuse or recycling of wastes can reduce the amount of fresh water and raw materials required for a process. While looking at the inputs to unit operations, think about the opportunities for reusing and recycling outputs from other operations.

Steps 4, 5, and 6 Summary

- By the end of step 6, you should have quantified all your process inputs.

- The net input of raw materials and water to the process should be established, having taken into account any losses incurred at the storage and transfer stages.

- Any reused or recycled inputs should be documented.

- All notes regarding raw material handling, process layout, water losses, or obvious areas where problems exist should be documented for consideration in phase 3.

Step 7. Quantifying Process Outputs

To calculate the second half of the material balance, the outputs from unit operations and the process as a whole need to be quantified.

Outputs include primary product, by-products, wastewater, gaseous wastes (emissions to atmosphere), liquid and solid wastes that need to be stored or sent off-site for disposal, and reusable or recyclable wastes. You may find that a table along the lines of Table 5–6 will help organize the input information. It is important to identify the units of measurement.

Table 5–6 Example of How to Tabulate Process Outputs

Unit Operation	Product	By-Product	Waste to Be Reused	Wastewater	Gaseous Emissions	Stored Wastes	Liquid/Solid Wastes Off-Site
A							
B							
C							
Total							

The assessment of the amount of primary product or useful product is a key factor in process or unit operation efficiency. If the product is sent off-site for sale, then the amount produced is likely to be documented in company records. However, if the product is an intermediate to be put into another process or unit operation, then the output may not be so easy to quantify. Production rates have to be measured over a period of time. Similarly, the quantification of any by-products may require measurement.

Step 8. Accounting for Wastewater

On many sites, significant quantities of both clean and contaminated water are discharged to a sewer or watercourse. In many cases, this wastewater has environmental implications and incurs treatment costs. In addition, wastewater may wash out valuable unused raw materials from the process areas.

It is extremely important to know how much wastewater is going down the drain and what the wastewater contains. The wastewater flows, from each unit operation as well as from the process as a whole, need to be quantified, sampled, and analyzed.

Identify the effluent discharge points; that is, where does wastewater leave the site? Wastewater may go to an effluent treatment plant or directly to a public sewer or a watercourse. One factor often overlooked is the use of several discharge points—*it is important to identify the location, type, and size of all discharge flows.*

- Identify where flows from different unit operations or process areas contribute to the overall flow. In this way, it is possible to piece together the drainage network for your site. This can lead to startling discoveries of what goes where!

- Once the drainage system is understood, it is possible to design an appropriate sampling and flow measurement program to monitor the wastewater flows and strengths from each unit operation.

- Plan your monitoring program thoroughly and try to take samples over a range of operating conditions, such as full production, startup, shutdown, and washing out. In the case of combined storm-water and wastewater drainage systems, ensure that sampling and flow measurements are carried out in dry weather.

- For small or batch wastewater flows it may be physically possible to collect all the flow for measurement using a pail and wristwatch. Larger or continuous wastewater flows can be assessed using flow measurement techniques.

The sum of the wastewater generated from each unit operation should be approximately the same as that input to the process. As indicated in step 6, note that double-counting can occur where wastewater is reused. This emphasizes the importance of understanding your unit operation and their interrelationships.

The wastewater should be analyzed to determine the concentration of contaminants. You should include wastewater analyses, such as

- pH.
- Chemical oxygen demand (COD).
- Biochemical oxygen demand (BOD_5).
- Suspended solids.
- Grease and oil.

Other parameters that should be measured depend on the raw material inputs. For example, an electroplating process is likely to use nickel and chromium. The metal concentrations of the wastewater should be measured to ensure that the concentrations do not exceed discharge regulations, but also to ensure that raw materials are not being lost to the drain. All toxic substances used in the process should be measured.

Take samples for laboratory analysis. Composite samples should be taken for continuously running wastewater. For example, a small volume, 100 ml, may be collected every hour through a production period of 10 hours to gain a 1 liter composite sample. The composite sample represents the average wastewater conditions over that time. Where significant flow variations occur during the discharge period, consideration should be given to varying the size of individual samples in proportion to flow rate to ensure that a representative composite sample is obtained. For batch tanks and periodic draindown, a single spot sample may be adequate (check for variations between batches before deciding on the appropriate sampling method).

Wastewater flows and concentrations should be tabulated as in Table 5–7.

Step 9. Accounting for Gaseous Emissions

To arrive at an accurate material balance, some quantification of gaseous emissions associated with the process is necessary. It is important to consider the actual and potential gaseous emissions associated with each unit operation, from raw material storage through to product storage.

Gaseous emissions are not always obvious and can be difficult to measure. Where quantification is impossible, estimations can be made using stoichiometric information. The following example illustrates the use of indirect estimation.

Consider coal burning in a boiler house. The assessor may not be able to measure the mass of sulfur dioxide leaving the boiler stack due to problems of access and lack of suitable sampling ports on the stack. The only information available is that the coal is of soft quality containing 3% sulfur by weight and, on average, 1000 kg of coal is burned each day.

First calculate the amount of sulfur burned:

1000 kg coal $\times$ 0.03 kg sulfur/kg coal = 30 kg sulfur/day.

The combustion reaction is approximately

$S + O_2 = SO_2.$

The number of moles of sulfur burned equal the number of moles of sulfur dioxide produced. The atomic weight of sulfur is 32 and the molecular weight of sulfur dioxide is 64. Therefore,

kg-moles S = 30 kg/32 kg per kg-mole
= kg-mole of SO_2 formed.

kg SO_2 formed = (64 kg SO_2/kg-mole) $\times$ kg-moles SO_2
= 64 $\times$ 30/32 = 60 kg.

Table 5-7 Example of How to Tabulate Wastewater Flows

Source	Discharge To								
	Public Sewer		Storm-Water Drain		Reuse		Storage		Total Wastewater Output
	Flow	Conc.	Flow	Conc.	Flow	Conc.	Flow	Conc.	Flow	Conc.
Unit operation A										
Unit operation B										
Unit operation C										

Note: Conc. is concentration.

Thus, it may be estimated that an emission of 60 kg sulfur dioxide takes place each day from the boiler stack.

- Record the quantified emission data in tabular form and indicate which figures are estimates and which are actual measurements.
- The assessor should consider qualitative characteristics at the same time as quantifying gaseous wastes.
  - Are odors associated with a unit operation?
  - Are there certain times when gaseous emissions are more prominent; are they linked to temperature?
  - Is any pollution control equipment in place?
  - Are gaseous emissions from confined spaces (including fugitive emissions) vented to the outside?
  - If gas scrubbing is practiced, what is done with the spent scrubber solution? Could it be converted to a useful product?
  - Do employees wear protective clothing, such as masks?

Step 10. Accounting for Off-Site Wastes

Your process may produce wastes that cannot be treated on-site. These need to be transported off-site for treatment and disposal. Wastes of this type are usually nonaqueous liquids, sludges, or solids. Often, wastes for off-site disposal are costly to transport and to treat. Therefore, *minimization of these wastes yields direct cost benefits.*

Measure the quantity and note the composition of any wastes associated with your process which need to be sent for off-site disposal. Record your results in a table like Table 5–8.

Ask several questions during the data collection stage:

- Where does the waste originate?
- Could the manufacturing operations be optimized to produce less waste?
- Could alternative raw materials be used that would produce less waste?

Table 5–8 Tabulating Wastes for Off-Site Disposal

	Liquid		Sludge		Solid	
Unit Operation	Quantity	Composition	Quantity	Composition	Quantity	Composition
A						
B						
C						

Note: Quantities in m³/annum or tonnes/annum.

- Is there a particular component that renders the whole waste hazardous and could this component be isolated?
- Does the waste contain valuable materials?
- Wastes for off-site disposal need to be stored on-site prior to dispatch. Does storage of these wastes cause additional emission problems? For example, are solvent wastes stored in closed tanks?
- How long are wastes stored on-site?
- Are stockpiles of solid waste secure or are dust storms a regular occurrence?

Steps 7, 8, 9, and 10 Summary

- At the end of step 10, the pollution prevention assessment team should have collated all the information required for evaluating a material balance for each unit operation and for the whole process.
- All actual and potential wastes should be quantified. Where direct measurement is impossible, estimates based on stoichiometric information should be made.
- The data should be arranged in clear tables with standardized units. Throughout the data collection phase, the assessors should make notes regarding actions, procedures, and operations that could be improved.

Step 11. Assembling Input and Output Information for Unit Operations

The total of what goes into a process must equal the total of what comes out. Prepare a material balance at a scale appropriate for the level of detail required in your study. For example, you may require a material balance for each unit operation or one for the whole process may suffice.

Preparing a material balance is designed to gain a better understanding of the inputs and outputs, especially waste, of a unit operation such that areas where information is inaccurate or lacking can be identified. Imbalances require further investigation. Do not expect a perfect balance—your initial balance should be considered a rough assessment to be refined and improved.

- Assemble the input and output information for each unit operation, then decide whether all the inputs and outputs need to be included in the material balance. For example, this is not essential where the cooling water input to a unit operation equals the cooling water output.
- Standardize units of measurement (liters, tons, or kilograms) on a per day, per year, or per batch basis.
- Summarize the measured values in standard units by reference to your process flow diagram. It may have been necessary to modify your process flow diagram following the in-depth study of the plant.

Step 12. Deriving a Preliminary Material Balance for Unit Operations

Now, it is possible to complete a preliminary material balance. For each unit operation, utilize the data developed in steps 1–10 and construct the material balance. Display your information clearly. Table 5-9 shows one way of presenting the material balance information.

Note that a material balance often must be carried out in weight units, since volumes are not always conserved. Where volume measurements have to be converted to weight units, take account of the density of the liquid, gas, or solid concerned.

Table 5–9 Preliminary Material Balance for Each Unit Operation

Unit Process A	Amounts (standard units per annum)
Inputs:	
Raw material 1	
Raw material 2	
Raw material 3	
Waste reuse	
Water	
Total	
Outputs:	
Product	
By-product	
Raw material storage and handling losses	
Reused wastes	
Wastewater	
Gaseous emissions	
Stored wastes	
Hazardous liquid waste transported off-site	
Hazardous solid waste transported off-site	
Nonhazardous liquid waste transported off-site	
Nonhazardous solid waste transported off-site	
Total	

Once the material balance for each unit operation has been completed for raw material inputs and waste outputs, it might be worthwhile repeating the procedure with respect to each contaminant of concern. It is highly desirable to carry out a water balance for all water inputs and outputs to and from unit operations, because water imbalances may indicate serious underlying process problems such as leaks or spills. The individual material balances may be summed to give a balance for the whole process, a production area, or a factory.

Step 13. Evaluating the Material Balance

The individual and sum totals making up the material balance should be reviewed to determine information gaps and inaccuracies. If you do have a significant material imbalance then further investigation is needed. For example, if outputs are less than inputs look for potential losses or waste discharges (such as evaporation). Outputs may appear greater than inputs if large measurement or estimating errors are made or some inputs have been overlooked.

At this stage you should take time to reexamine the unit operations to attempt to identify where unnoticed losses may be occurring. It may be necessary to repeat some data collection activities.

Remember that you need to be thorough and consistent to obtain a satisfactory material balance. The material balance not only reflects the adequacy of your data collection but, by its very nature, ensures that you have a sound understanding of the processes involved.

Step 14. Refining the Material Balance

Now, you can reconsider the material balance equation by adding those additional factors identified in the previous step. If necessary, estimates of unaccountable losses have to be calculated.

Note that, in the case of relatively simple manufacturing plants, preparation of a preliminary material balance and its refinement (steps 13 and 14) can be combined usefully. For more complex pollution prevention assessments, however, two separate steps are likely to be more appropriate.

Remember, ideally, the inputs should equal the outputs, but in practice, this rarely is the case and some judgment is required to determine what level of accuracy is acceptable.

In the case of high-strength or hazardous wastes, accurate measurements are needed to design waste reduction options. The material balance for a number of unit operations might need to be repeated. Again, continue to review, refine, and where necessary, expand your database. The compilation of accurate and comprehensive data is essential for a successful pollution prevention assessment and subsequent waste reduction action plan. *You cannot reduce what you do not know is there.*

Steps 11, 12, 13, and 14 Summary

By the end of step 14, you should have assembled information covering process inputs and process outputs. These data should be organized and presented clearly in the form of material balances for each unit operation. These data form the basis for the development of an action plan for waste minimization.

Phase 3. Synthesis

Phases 1 and 2 covered planning and undertaking a pollution prevention assessment, resulting in the preparation of a material balance for each unit operation.

Phase 3 represents the interpretation of the material balance to identify process areas or components of concern.

The material balance focuses on the attention of the assessor. The arrangement of the input and output data in the form of a material balance facilitates your understanding of how materials flow through a production process.

To interpret a material balance, it is necessary to understand the normal operating performance. How can you assess whether a unit operation is working efficiently if you do not know what is normal? A member of your team must have a good working knowledge of the process.

To a trained eye, the material balance indicates areas for concern and helps prioritize problem wastes.

You should use the material balance to identify the major sources of waste, look for deviations from the norm in terms of waste production, identify areas of unexplained losses, and pinpoint operations that contribute to flows that exceed national or site discharge regulations. Different waste reduction measures require varying degrees of effort, time, and financial resources. They can be categorized as two groups:

Group 1. Obvious waste reduction measures, including improvements in management techniques and housekeeping procedures that can be implemented cheaply and quickly; that is, low cost/no cost.

Group 2. Long-term reduction measures, involving process modifications or process substitutions to eliminate problem wastes.

Increased reuse and recycling to reduce waste falls between the immediate and the more substantial waste reduction measures.

Conducting Cleaner Production Audits

Steps 15, 16, and 17 describe how to identify waste reduction measures.

Step 15. Examining Obvious Waste Reduction Measures

It may have been possible to implement very obvious waste reduction measures already, before embarking on obtaining a material balance (refer back to step 3). Now, consider the material balance information in conjunction with visual observations made during the whole of the data collection period to pinpoint areas or operations where simple adjustments in procedure could greatly improve the efficiency of the process by reducing unnecessary losses.

Use the information gathered for each unit operation to develop better operating practices for all units.

Significant waste reductions can often be achieved by improved operation, better handling, and generally taking more care. The following list of waste reduction hints can be implemented immediately with no or only small extra costs.

Specifying and Ordering Materials

- Do not overorder materials especially if the raw materials or components can spoil or are difficult to store.
- Try to purchase raw materials in a form that is easy to handle, for example, pellets instead of powders.
- It is often more efficient and certainly cheaper to buy in bulk.

Receiving Materials

- Demand quality control from suppliers by refusing damaged, leaking, or unlabeled containers.
- Undertake a visual inspection of all materials coming onto the site.
- Check that a sack weighs what is should weigh and that the volume ordered is the volume supplied.
- Check that composition and quality are correct.

Material Storage
- Install high-level control on bulk tanks to avoid overflows.
- Bund tanks to contain spillages.
- Use tanks that can be pitched and elevated, with rounded edges for ease of draining and rinsing.
- Dedicated tanks, receiving only one type of material, do not need to be washed out as often as tanks receiving a range of materials.
- Make sure that drums are stored in a stable arrangement to avoid damage while in storage.
- Implement a tank checking procedure—dip tanks regularly and document to avoid discharging a material into the wrong tank.
- Evaporation losses are reduced by using covered or closed tanks.

Material and Water Transfer and Handling
- Minimize the number of times materials are moved on-site.
- Check transfer lines for spills and leaks.
- Is flexible pipework too long?
- Catch draining from transfer hoses.
- Plug leaks and fit flow restrictions to reduce excess water consumption.

Process Control
- Design a monitoring program to check the emissions and wastes from each unit operation.
- Regular maintenance of all equipment helps reduce fugitive process losses.
- Feedback on how waste reduction is improving the process motivates the operators—it is vital that employees are informed of why actions are taken and what it is hoped to achieve.

Cleaning Procedures

- Minimize the amount of water used to wash out and rinse vessels. On many sites, indiscriminate water use contributes a large amount to wastewater flows. Ensure that hoses are not left running by fitting self-sealing valves.
- Investigate how washing water can be contained and used again before discharge to drains. The same applies to solvents used to clean; these can often be used more than once.

Tightening up housekeeping procedures can reduce waste considerably. Simple, quick adjustments should be made to your process to achieve a rapid improvement in process efficiency. Where such obvious reduction measures do not solve the entire waste disposal problem, more detailed consideration of waste reduction options are needed (steps 16–18).

Step 16. Targeting and Characterizing Problem Wastes

Use the material balance for each unit operation to pinpoint the problem areas associated with your process.

The material balance exercise may have brought to light the origin of wastes with high treatment costs or may indicate which wastes are causing process problems in which operations. The material balance should be used for your priorities for long-term waste reduction.

At this stage, it may be worthwhile considering the underlying causes as to why wastes are generated and the factors that lead to these causes; for example, poor technology, lack of maintenance, or noncompliance with company procedures.

Additional sampling and characterization of your wastes might be necessary, involving more in-depth analysis to ascertain the exact concentrations of contaminants.

List the wastes in order of priority for reduction actions.

Step 17. Segregation

Segregation of wastes can offer enhanced opportunities for recycling and reuse with resultant savings in raw material costs. Concentrated simple wastes are more likely to be of value than dilute or complex wastes.

Mixing wastes can enhance pollution problems. If a highly concentrated waste is mixed with a large quantity of weak, relatively uncontaminated effluent, the result is a larger volume of waste requiring treatment. Isolating the concentrated waste from the weaker waste can reduce treatment costs. The concentrated waste could be recycled or reused or may require physical, chemical, and biological treatment to comply with discharge consent levels, whereas the weaker effluent could be reused or may require only settlement before discharge. Therefore, waste segregation can provide more scope for recycling and reuse while at the same time reducing treatment costs.

Review your waste collection and storage facilities to determine if waste segregation is possible. Adjust your list of priority wastes accordingly.

Step 18. Developing Long-Term Waste Reduction Options

Waste problems that cannot be solved by simple procedural adjustments or improvements in housekeeping practices require more substantial long-term changes. It is necessary to develop possible prevention options for the waste problems.

The process or production changes that may increase production efficiency and reduce waste generation include

- Changes in the production process, continuous versus batch.
- Equipment and installation changes.
- Changes in process control, automation.
- Changes in process conditions, such as retention times, temperatures, agitation, pressure, catalysts.
- Use of dispersants in place of organic solvents where appropriate.
- Reduction in the quantity or type of raw materials used in production.
- Raw material substitution through the use of wastes as raw materials or the use of different raw materials that produce less waste or less hazardous waste.
- Process substitution with cleaner technology.

Waste reuse often can be implemented if materials of sufficient purity can be concentrated or purified. Technologies such as reverse osmosis, ultrafiltration, electrodialysis, distillation, electrolysis, and ion exchange may enable materials to be reused and reduce or eliminate the need for waste treatment.

Where waste treatment is necessary, a variety of technologies should be considered. These include physical, chemical, and biological treatment processes. In some cases the treatment method can also recover valuable materials for reuse. *Another industry or factory may be able to use or treat a waste that you cannot treat on-site. It may be worth investigating the possibility of setting up a waste exchange bureau as a structure for sharing treatment and reuse facilities.*

Consider also the possibilities for product improvements or changes yielding cleaner, more environmentally-friendly products, both for existing products and in the development of new products.

Steps 15, 16, 17, and 18 Summary
At the end of step 18 you should have identified all the waste reduction options that can be implemented.

Step 19. Environmental and Economic Evaluation of Waste Reduction Options
To decide which options should be developed to formulate a waste reduction action plan, each option should be considered in terms of environmental and economic benefits.

Environmental Evaluation
It is often taken for granted that reduction of a waste has environmental benefits. This is generally true; however, there are exceptions to the rule. For example, reducing one waste may give rise to pH imbalances or produce another that is more difficult to treat, resulting in a net environmental disadvantage.

In many cases, the benefits may be obvious, such as the result of the removal of a toxic element from an aqueous effluent by segregating the polluted waste or changing the process in such a way that the waste is prevented.

In other cases, the environmental benefits may be less tangible. Creating a cleaner, healthier workplace increases production efficiency but this may be difficult to quantify.

For each option a series of questions should be asked:

- What is the effect of each option on the volume and degree of contamination of process wastes?
- Does a waste reduction option have cross-media effects? For example, does the reduction of a gaseous waste produce a liquid waste?
- Does the option change the toxicity, degradability, or treatability of the wastes?
- Does the option use more or less nonrenewable resources?
- Does the option use less energy?

Economic Evaluation

A comparative economic analysis of the waste reduction options and the existing situation should be undertaken. Where benefits or changes cannot be quantified (e.g., reduction in future liability, worker health and safety costs), some form of qualitative assessment should be made.

Economic evaluations of waste reduction options should involve a comparison of operating costs to illustrate where cost savings would be made. For example, a waste reduction measure that reduces the amount of raw material lost to the drain during the process results in reduced raw material costs. Raw material substitution or process changes may reduce the amount of solid waste that has to be transported off-site. Therefore, the transport costs for waste disposal are reduced.

In many cases, it is appropriate to compare the waste treatment costs under existing conditions with those associated with the waste reduction option.

The size of treatment plant and the treatment processes required may be altered significantly by the implementation of waste reduction options. This should be considered in an economic evaluation.

Calculate the annual operating costs for the existing process, indicating waste treatment, and estimate how these would be altered with the introduction of waste reduction options. Tabulate and compare the process and waste treatment operating costs for both the existing and proposed future waste management options. The example given in Table 5–10 shows the typical cost components. In addition, if there are any monetary benefits (e.g., recycled or reused materials or wastes), then these should be subtracted from the total process or waste treatment costs as appropriate.

Now that you have determined the likely savings in terms of annual process and waste treatment operating costs associated with

Table 5–10 Example of Summarizing Annual Process and Waste Treatment Operating Costs

	Annual Cost
Process operating costs:	
Raw material 1	
Raw material 2	
Water	
Energy	
Labor	
Maintenance	
Administration	
Other	
Total	
Waste treatment operating costs:	
Raw material (e.g., lime)	
Raw material (e.g., flocculent)	
Water	
Energy	
Trade effluent discharge costs	
Transportation	
Off-site disposal	
Labor	
Maintenance	
Administration	
Other (e.g., violation, fines)	
Total	

each option, consider the necessary investment required to implement each option. Investment can be assessed by looking at the payback period for each option. The payback period is the time taken for a project to recover its financial outlay. A more-detailed investment analysis may involve an assessment of the internal rate of return (IRR) and net present value (NPV) of the investment based on discounted cash flows. Analysis of the investment risk allows you to rank options. Consider the environmental benefits and the savings in process and waste treatment operating costs along with the payback period for an investment, to decide which options are viable.

Step 20. Developing and Implementing an Action Plan: Reducing Wastes and Increasing Production Efficiency

Consider the immediate reduction measures identified in step 15 along with the long-term waste reduction measures that have been evaluated in steps 18 and 19. These measures should form the basis of the waste reduction action plan. Discuss your findings with members of staff and develop a workable action plan.

Prepare the ground for the waste reduction action plan. Its implementation should be preceded by an explanation of the objectives behind undertaking a CP/P2 assessment.

It is necessary to convince those who must work with new procedures that the change in philosophy from end-of-pipe treatment to waste prevention makes sense and improves efficiency.

Use posters around the site to emphasize the importance of waste reduction to minimize production and waste treatment or disposal costs, and where appropriate, improve the health and safety of company personnel.

Set out the intended action plan within an appropriate schedule. Remember it may take time for the staff to feel comfortable with a new way of thinking. Therefore, it is a good idea to implement waste reduction measures slowly but consistently to allow everyone time to adapt to these changes.

Set up a monitoring program to run alongside the waste reduction action plan, so that actual improvements in process efficiency can be measured. Relay these results back to the workforce as evidence of the benefits of waste reduction. Adopt an internal record-keeping system for maintaining and managing data to support material balances and waste reduction assessments.

It is likely that you will have highlighted significant information gaps or inconsistencies during the pollution prevention assessment investigations. You should concentrate on these gaps and explore ways of developing the additional data. Is outside help required?

A good way of providing waste reduction incentives is to set up an internal waste charging system, those processes that create waste in great volume or that are difficult and expensive to handle have to contribute to the treatment costs on a proportional basis. Another method of motivating staff is to offer financial rewards for individual waste-saving efforts, drawing on the savings gained from implementing waste reduction measures.

Pollution prevention assessments should be a regular event. Attempt to develop a specific pollution prevention assessment approach for your own situation, keeping abreast of technological advances that could lead to waste reduction and the development of "cleaner" products. Train process employees to undertake material balance exercises.

Training people who work on the process to undertake a pollution prevention assessment helps raise awareness in the workforce. Without the support of the operators, waste reduction actions will be ineffectual—these are the people who can really make a difference to process performance.

Step 20 Summary

- Prepare the ground for the P2 action plan, ensuring that support for the assessment and implementation of the results is gained from senior management. Implement the plan slowly to allow the workforce to adjust.
- Monitor process efficiency.
- Relay results back to the workforce to show them the direct benefits.

Toolkit and Worksheets

The following worksheets will assist in conducting a CP/P2 assessment. The worksheets are contained in the toolkit CD.

The CD P2 toolkit runs on a Microsoft Access platform. You will need Microsoft Access to use the toolkit. It is designed to provide a simple database management tool that will facilitate a P2 audit. The CD allows multiple files to be stored, retrieved, and printed out in a simplified reporting format that will assist the team in evaluating multiple P2 opportunities.

Worksheet 1   The Pollution Prevention Team

COMPLETED BY: DATE:

COMPANY: FACILITY NAME:

TEAM LEADER: TITLE: PHONE:

Responsibilities

Member Names	Corporate	Team	Phone number
_____	Plant superintendent Quality control	Leader Engineering—process	_____
_____	Environmental consultant Sales/accounting	Environment Cost—analyses	_____

Responsibility of pollution prevention team:

1. _____

2. _____

    _____

Worksheet 2

SITE:

COMPLETED BY: DATE:

1. Facility name: _____

2. Facility/corporate owner: _____

3. Contact: _____

4. Site address: _____
    Street: _____ City: _____
    State, Postal Code: _____

5. Legal description of site location: _____
    _____

6. Estimated site size: _____

7. Mailing address: _____
    Street: _____ City: _____
    State, Postal Code: _____

8. Phone and fax numbers: _____
    Phone: _____ Fax: _____ Other: _____

Worksheet 3 Environmental Site-Specific Factors

COMPLETED BY: _____ DATE: _____

1. Distance to nearest fish-bearing water body: _____
   a. Name of water body: _____

2. Distance to other ecologically-sensitive areas (e.g., waterfowl breeding, other):

   Distance   Area Description
   _____
   _____

3. Distance to nearest populated area: _____
   Name of municipality: _____
   Distance to nearest groundwater well: _____
   Depth to groundwater at site: _____
   Composition of site soil: _____

4. Average annual precipitation at site: _____
   (List data source): _____

5. Is site in a 100-year flood plain? _____
   (List data source): _____

6. Release points of storm-water discharges:
   _____
   _____
   _____
   _____

Worksheet 4   Preaudit Information Collection

COMPLETED BY: DATE:

1. Facility name: _____

2. Contact person for data: _____ Phone: _____

3. Chemicals in use: _____

4. Summary of essential documentation for site audit:

Document	Available (Y/N)	Date of Latest Version	Location/ Comments
Company literature (e.g., products, services, other)			
Process descriptions			
Diagrams, blueprints, drawings of buildings, process areas, storage areas, other			
Design-related information (equipment lists, equipment specifications, process flow diagrams)			
Operating manuals and standard operating procedures (SOPs)			
Inventory of chemicals			
Product inventory			
Material safety data sheets			
Pollution monitoring data/reports			
Hazardous waste manifests			
Environmental audit reports			
Regulatory permits and correspondence			
Fire inspection reports			
Employee training records			
Operator data logs			
Preventive maintenance logs			

Conducting Cleaner Production Audits

Worksheet 4 *continued*

5. Available auxiliary information:

Document	Available (Y/N)	Date of Latest Version	Location/ Comments
Material balance analyses information: Input streams			
Product streams			
Waste streams			
Energy and water use			
Fuel consumption			
Electricity records			
Labor usage costs			
Operating and maintenance costs			
Water usage costs			

Worksheet 5   Chemical Use Data Summary

COMPLETED BY: DATE:

1. Facility name:_____

2. Materials summary:

	A	B	C	D	E	F	G
Raw material:							
Trade name							
Chemical name							
Components/ingredients/ attributes of concern							
Components and concentration (Specify Units—%, PPM, other)							
Annual consumption rate (Specify units—lb, kg, ton, …)							
Suppliers:							
Supplier 1							
Supplier 2							
Costs:							
Purchase price per unit, specify unit purchase ($/unit value)							
Overall annual cost ($)							
Shipping and storage information:							
Delivery mode (type vehicle)							
Normal order/delivery size							
No. shipments per year							
Shipping container (size and type)							
Storage mode							
Transfer mode							
Inventory size (max.)							
Empty container management (describe):							

Worksheet 6 Regulatory Compliance

COMPLETED BY: DATE:

1. Facility name: _____
2. Contact person for data: _____ Phone: _____
3. Air emissions and controls
   (a) Check sources of controlled emissions:
   - ☐ Vacuums and kilns
   - ☐ Reactors
   - ☐ Boilers
   - ☐ Solvent use stations
   - ☐ Degreasers
   - ☐ Tank vents
   - ☐ Bulk fuel handling
   - ☐ Incinerators
   - ☐ Mixers and blending
   - ☐ Other, specify

   (b) List permits and/or relevant regulations responsible agency:

   _____
   _____

   (c) Check the appropriate answer:

	No	Yes	NA
Is pollution control equipment operational?	☐	☐	☐
Are monitoring results available or in a report?	☐	☐	☐
Is the monitoring frequency and analytical sensitivity in accordance with permits and other regulatory requirements?	☐	☐	☐
Do the results indicate compliance with regulatory limits within the last year?	☐	☐	☐
Do the results indicate compliance with regulatory limits within the last three years?	☐	☐	☐
Has the necessary reporting been made to the regulatory agencies?	☐	☐	☐

4. Liquid effluent releases:
   (a) Identify sources of effluent releases:

Types	Discharge Location
Storm-water runoff	
Treatment system discharge	
Other	

Worksheet 6 *continued*

(b) List permits and/or regulatory agency responsible:

_____

_____

(c) Check the appropriate answer: *No Yes NA*

	No	Yes	NA
Is pollution control equipment operational?	☐	☐	☐
Are monitoring results available or in a report?	☐	☐	☐
Is the monitoring frequency and analytical sensitivity in accordance with permits and other regulatory requirements?	☐	☐	☐
Do the results indicate compliance with regulatory limits within the last year?	☐	☐	☐
Have activated carbon beds been changed?	☐	☐	☐
Have filtration units been changed?	☐	☐	☐

Conducting Cleaner Production Audits

Worksheet 7 Control of Bulk Liquids and Gases

COMPLETED BY: DATE:

1. Inventory of bulk liquid and gases:

Product	Storage type (UST or AST*) and volume (specify units)	Is containment adequate? See 2(c)
A.		
B.		
C.		
D.		
E.		
F.		
G.		

*Underground storage tank (UST); aboveground storage tank (AST).

2. Aboveground tanks

   Check the appropriate answer: No Yes NA

   Are tanks and drums placed on top of an
   impermeable floor?   ☐   ☐   ☐

   Is the floor sealed and there are no signs of
   cracks?   ☐   ☐   ☐

   Are all containment volumes in excess of 150%
   of the largest tank or the volume of the largest
   tank?   ☐   ☐   ☐

   Are tanks free of corrosion and physical damage?   ☐   ☐   ☐

   Are all tanks securely mounted and protected
   from vehicle impact?   ☐   ☐   ☐

3. Underground storage tanks (USTs)

   Check the appropriate answer: No Yes NA

   Are all USTs less than 10 years old?   ☐   ☐   ☐

   Are the USTs and piping leak tested annually
   and are records maintained?   ☐   ☐   ☐

Worksheet 7 *continued*

4. All tanks and drums

   Check the appropriate answer: No Yes NA

   Are tanks and drums secure from tampering and
   accidental impacts?   ☐   ☐   ☐

   Do tanks have overfill protection and/or alarms?   ☐   ☐   ☐

   Are bulk drains isolated from the chemical
   unloading area and bulk tank storage area?   ☐   ☐   ☐

   Are nongravity-feed systems used for fuel supply
   and dispensing?   ☐   ☐   ☐

   Are dispensing systems deenergized and locked
   when not in use?   ☐   ☐   ☐

   Are there drip and spill containment provisions at
   chemical dispensing areas?   ☐   ☐   ☐

   Are there spill-response kits provided at bulk
   chemical storage and handling stations?   ☐   ☐   ☐

   Is there a written spill-response program?   ☐   ☐   ☐

   Has spill-response training been provided
   within the last year?   ☐   ☐   ☐

5. Firefighting

   Check the appropriate answer: No Yes NA

   Is firefighting equipment readily available?   ☐   ☐   ☐

   Is there a written fire response and control plan?   ☐   ☐   ☐

   Have employees been trained within the last
   year?   ☐   ☐   ☐

Worksheet 8 Facility Status Priorities Identification

COMPLETED BY: DATE:

1. Summary of regulatory compliance:

Regulatory Permit	Permit No.	Date Issued	Renewal Date	Compliance Status
Liquid effluent wastes				
Air permit				
Solid waste permit				
Other (List here and below)				

   Review data from previous worksheets and list priority items that should be addressed.

2. Summary of areas of concern, such as catastrophic releases and site contamination:

List no. and worksheet ID	P2 team priority and basis of recommendation

Worksheet 9   Identification of P2 Measures

COMPLETED BY: DATE:

1. Recommended measures for immediate implementation (Low-cost/No-cost measures):

    Recommended Measure   Priority   Type of P2 Measure

2. Identification of P2 opportunities and possible options:

    Pollution Prevention Opportunity   Possible Option

Worksheet 10 CP/P2 Technical Feasibility Assessment

COMPLETED BY: _____   DATE: _____

Apply this form for each P2 opportunity under assessment.

1. Describe area of concern (e.g., pollution, waste, material losses, excess water, energy, safety, other): _____
2. CP/P2 opportunity description: _____
3. Type CP/P2 option (check):
   - ☐ Equipment related
   - ☐ Process related
   - ☐ Raw material related
   - ☐ Personnel/safety related
   - ☐ Energy related
4. Describe possible personnel requirements (e.g., training, safety, other). _____
5. Provide description of space and utility requirements. Are they available? _____
6. Explain whether production quality or services will be affected? _____
7. Does proposal create other environmental or health and safety problems? If yes, explain. _____
8. Technical feasibility ranking (check):
   - ☐ Easily achievable
   - ☐ Major process change needed
   - ☐ Minor equipment/piping changes
   - ☐ Requires process development
   - ☐ Minor process changes
   - ☐ Requires new process
9. Explain whether the technical feasibility of the opportunity requires economic analysis. _____

Worksheet 11 Part 1: P2 Financial Attractiveness

COMPLETED BY: DATE:

Apply this form for each P2 opportunity under assessment.

1. Description of CP/P2 option:
   Procurement Expenses   *Costs*

2. Capital costs:
   Recycle equipment _____
   Materials (piping, pumps, fans, blowers, etc.) _____
   Site prep and installation _____
   Engineering services _____
   Permitting _____
   Utilities connection _____
   Total capital costs _____

3. Annual operating costs:
   Interest expense (Capital cost × interest) _____
   Depreciation expense _____
   Initial training costs _____
   Operating expenses (labor, utilities, maintenance) _____
   Environmental transformation costs/waste disposal _____
   Monitoring and other environmental monitoring costs _____
   Total—Year 1 _____
   Interest expense _____
   Depreciation expenses _____
   Operating expenses (use as basis 5% per year increase or state otherwise) _____
   Environmental transformation costs/waste disposal (use as basis 5% per year increase or state otherwise) _____
   Monitoring and other environmental monitoring costs (use as basis 5% per year increase or state otherwise) _____
   Total—Year 1 _____

Worksheet 11 Part 2: Assessment of Apparent Cost Savings

COMPLETED BY: DATE:

	Quantity (specify units)	Cost ($)	Savings ($)	Comments
Product recovery:				
Amount of recovered products				
Costs of recovered products				
Total savings for recovered product				
Reclaimed water:				
Amount of reclaimed water				
Cost per volume				
Opportunity cost (value of reclaimed water for higher-value applications)				
Total savings of reclaimed water (cost per vol. × amt. reclaimed + opportunity costs)				
Pollution fees savings:				
Savings in reduced permitting				
Savings in reduced monitoring				
Savings in reduced sewer discharge fees				
Savings in reduced pollution fees (air + water other than sewer effluent)				

Worksheet 11 Part 2 *continued*

	Quantity (specify units)	Cost ($)	Savings ($)	Comments
Pollution fees savings (cont.):				
Savings from solid waste disposal				
Total pollution fees savings				
Energy savings:				
Annual electricity				
Fuel oil savings				
Steam reduction savings				
Total energy savings				
Labor savings:				
Amount of reduced labor (e.g., hours)				
Cost of labor per unit				
Total labor savings				
Total apparent savings				
5. Cost-benefit savings				
Real cost savings (total annual costs − total apparent costs)				

6
IER Case Study

Introduction

This chapter provides a case study where elements of an integrated EMS/P2 program were placed in a refinery in Jordan (Jordan Petroleum Refinery Co.). The activity was designed and managed by me under The Reuse for Industry, Agriculture, and Landscaping (RIAL) Project funded by the United States Agency for International Development (USAID) and implemented under a contract managed by CDM International.

The role of RIAL is to assist the government of Jordan (GOJ) to further develop the sustainability of water reuse. One of the major components of RIAL is Task 3 (Water Reuse, Conservation, and Pollution Control in Industries), which focuses on developing a national strategy for the enhancement of industrial water recycling, reuse, and conservation.

Under Task 3.2 (Implementation of Integrated EMS/P2 Programs), RIAL focuses on the application of pollution prevention (P2) through ISO 14001 for improving enterprise competitiveness through improved environmental performance. By reducing pollution, waste, and inefficiencies at the source, less energy and materials are consumed in manufacturing and less of a dependence on end-of-pipe treatment technologies is achieved. Task 3.2 focuses on enterprise-specific programs that build elements of ISO 14001 and P2 into Jordanian business practices. By demonstrating to enterprises that good environmental performance leads to improved financial performance, enterprises become committed to reducing waste and inefficiencies at the source.

This case study helps illustrate that the application of an EMS can greatly improve both financial and environmental performance, helping make a company more competitive and sustainable. Major accomplishments achieved under this project were

- The establishment of a formal environmental program and corporate environmental policy statement.
- The formation of an EMS executive committee comprised of senior management.
- The formation of an EMS/P2 management team. On-site training of technical staff and senior management on P2, EMS, ISO 14001, EMIS (environmental management information systems), and preparing P2 cost-benefit analyses.
- The completion of a facilitywide IER (initial environmental review) and a ranking of the environmental aspects of the refinery.
- The completion of a P2 audit for the utilities section, in which no-cost/low-cost, moderate-cost, and high-cost P2 opportunities were identified and cost-benefit analyses prepared.
- The implementation of corrective actions.

Under this program, approximately $1.4 million in material savings alone were identified from P2 opportunities. When incremental savings in energy, labor productivity, and higher throughputs are considered, the actual savings to the enterprise exceeded $2.5 million based on similar exercises performed at other refineries. The P2 opportunities identified in the program further reduced water demand by about 250,000 m³/yr and achieved a total material savings of 278,000 ton/yr.

The Methodology Applied

The refinery has several operations, including a major fuel transport and loading facility in the port of Aqaba. Project resources were not sufficient to address the needs of the entire enterprise. Hence, the approach adopted was to transfer sufficient tools and knowledge that key refinery personnel would learn the essential elements of an integrated EMS/P2 program for one part of the refinery. This approach enabled a transfer of skills and a model EMS to be established within one portion of the refinery and, hence, within one part of the corporation. From there, the enterprise had sufficient experience, knowledge base, and tools to roll the model out on a corporatewide basis.

A stepwise model was created for implementing the program:

Step 1. *Meetings and follow-up discussions with senior-level corporate management.* In these, a formal commitment to the program is obtained, including a general definition and commitment to provide enterprise resources to the implementation of the program.

Step 2. *Initial training on P2 and EMS.* This training was later reinforced at several points in the program with more extensive training and on-site exercises, where enterprise staff learned how to perform an IER, rank the environmental aspects of their operation, and conduct a P2 audit.

Step 3. *A preaudit of the facility, in which a candidate process is selected for implementing a model EMS/P2 program.* This preaudit also defined the level of manpower and logistical support needed on the part of the enterprise and formed the basis for a detailed action plan and implementation schedule.

Step 4. *Corporate-level follow-up meeting and commitment.* A meeting was conducted with the senior corporate management, in which the program objectives and deliverables were defined, the level of support required on the part of the enterprise and costs discussed along with the potential payback and benefits from the program, and a formal implementation schedule reviewed. This meeting resulted in the creation of a memorandum of understanding (MOU) between the enterprise and RIAL.

Step 5. *Corporate environmental policy statement.* To ensure senior corporate commitment, the enterprise was required to create a corporate environmental policy statement. Senior management was provided examples of policy statements and guidance in its creation. The policy statement, however, is a product of the enterprise and reflects senior-level policy toward environmental performance and water management and a statement of core values on environmental protection. No such written policy existed at the refinery prior to the start of the program. The policy statement has since been published in local news media, in company literature distributed to employees and the enterprise's suppliers, and on the enterprise's Web site.

Step 6. *Establishing an EMS/P2 management team.* The team comprises key enterprise personnel with RIAL team members serving as technical advisors to the management team. At the start of the program, the refinery had one person working part-time on corporatewide environmental management. This individual's responsibilities extended to other facilities as well. The EMS/P2 management team is now a fully staffed department with four people working full-time. Additionally, the enterprise is currently reorganizing the Health and Safety Department and plans to integrate them into the EMS/P2 management team.

Step 7. *Conducting an IER.* The EMS/P2 management team, initially under the direction and guidance of Task 3 technical advisors, conducted an IER for the entire refinery. In conducting this exercise, an audit questionnaire was devised to help guide the team. This IER audit questionnaire was refined and adapted in the field to obtain more-detailed information on the process operation and define initial P2 opportunities.

Step 8. *Environmental matrix*. After data collection, a matrix was created listing all of the environmental aspects. Through a series of exercises and on-site training, the EMS/P2 team acquired the skills to rank the environmental aspects using a numerical scoring system described in the publication *Green Profits*.[1] This enabled the team to establish priorities for the environmental aspects in terms of public health risks, worker safety, interactions with the environment, financial and material losses, and manufacturing productivity. Additionally, a number of no-cost/low-cost P2 opportunities having immediate paybacks were identified. From this information, the team identified several areas for a focused P2 audit and an action plan for immediate corrective actions (CA) on the P2 opportunities.

Step 9. *CA action plan*. The CA action plan was presented to management, first and foremost, to strengthen the commitment to the program by demonstrating early on that P2 can result in financial savings and, second, to actually capture some savings. The majority of the no-cost/low-cost P2 opportunities identified fell into the housekeeping category.

Step 10. *P2 audit*. The EMS/P2 management team, with the assistance of RIAL Task 3 members, conducted a P2 audit following the 22-step procedure outlined in the publication *Green Profits* and summarized in Chapter 3. The steps were implemented over the three-phase system defined in the publication: phase I consisted of initial data gathering, including the development of process diagrams. Phase II comprised the quantification of process flows through the application of material and energy balances. Phase III, synthesis, comprised quantifying losses, calculating savings from source reduction or P2 alternatives, and preparing cost-benefit analyses. The P2 opportunities were organized into three cost categories: low cost/no cost, moderate cost, and high cost. The EMS/P2 management team

---

[1] N. P. Cheremisinoff and A. Bendavid-Val, *Green Profits: A Manager's Handbook to ISO 14001 and Pollution Prevention* (Oxford: Butterworth-Heinemann Publishers, 2001).

then prepared a formal report and presentation of the recommendations for senior management.

Step 11. *Conducting an exit interview.* A joint presentation by the refinery's EMS/P2 management team and task 3 members was made to the senior management. This meeting resulted in support for immediate corrective actions for the no-cost/low-cost P2 opportunities, a commitment to implement the moderate-cost P2 opportunities, and the formation of a project cost-estimating group to perform a more-detailed cost-benefit analysis of the high-cost opportunities.

Refinery History

The facility started operation in 1961 and presently employs 3,493 full-time personnel. Its mission is to supply the country with the petroleum products that meet a set of Jordanian standards, which are consistent with international specifications.

The enterprise is a public shareholding–limited company with a concession located in the Zarqa industrialized region. It is one of the major economic pillars in the kingdom and a large corporation by Jordanian standards.

The refinery receives its crude oil from Saudi Arabia and other miscellaneous suppliers based on open tenders for services and materials. The principle products are liquefied petroleum gas, gasoline (regular leaded, super leaded, and super unleaded), jet fuel, kerosene, diesel, fuel oil, asphalt, lubricating oils, and liquefied petroleum gas (LPG) cylinders. Major units of the refinery are

- Three crude distillation units.
- Two vacuum distillation units.
- Naphtha hydrotreating unit.
- Platforming unit.
- Fluid catalytic cracking unit.
- Hydrocracking unit.
- LPG recovery unit.

- Hydrogen plant.
- Two asphalt units.

Environmental Management System Structure

A management structure for an integrated EMS/P2 was established at the enterprise. This structure consists of

- An executive committee composed of senior management who oversee all environmental issues.
- An EMS management team that continually reviews policy and provides recommendations and guidance on corrective actions for improving environmental performance.
- A dedicated P2 team that reports monthly to the EMS management team. For expediency the EMS management team is referred to as the EMS/P2 management team by senior management. The P2 team is composed of technical experts from several departments. Their charter is to identify practices and technology investments that reduce waste, inefficiency, and pollution at the source. The team focuses on cost-effective recommendations and has been provided the tools to perform cost-benefit analyses that enable management to prioritize and make recommendations on P2 investments.

An overall management structure for an integrated EMS/P2 program, which incorporates the refinery support and operational departments into the continual improvement cycle of ISO 14001, was established as a basis for reorganization. The model presented in Chapter 2 was closely followed.

The Initial Environmental Review

The purpose of an IER is to establish a baseline on the environmental performance of the enterprise. Its goal is to identify the environmental aspects and their interactions with the facility, the public, the environment, and the business itself. It encompasses a review of environmental performance, regulatory compliance or lack thereof, and identifies negative interactions. Once a baseline performance level has been established, a basis for benchmarking improvements over time is possible and becomes a crucial aspect in the continual improvement cycle of an EMS.

The model approach to conducting an IER based on the publication *Green Profits* was applied at the enterprise. The IER was performed for the entire refinery.

The IER was started in early December 2004 by filling two detailed audit questionnaires. One was an industrial wastewater audit questionnaire; the second was the IER information collection form. The audit forms were later modified and adapted for more-detailed information gathering by the EMS/P2 management team.

The environmental aspects were identified and their impacts assessed from a semiqualitative standpoint. The environmental aspects were then ranked in accordance with their impacts on health risks, cost of managing to the enterprise based on current practices, resource utilization, and other factors. The IER also provided a baseline for the enterprise to begin tracking improvements and savings from the implementation of corrective actions.

The following discussions provide a summary of the environmental aspects along with initial P2 recommendations.

Crude Oil Unloading Station

Crude oil is imported from neighboring countries via 30–metric ton road tankers. The crude is unloaded to the crude tank pit through hoses and a transfer pumping station to storage tanks. Figure 6–1 shows a photograph of the unloading station.

The process of unloading is done manually. At the start of the unloading operation, the operator opens the valves to discharge accumulated water in the heal of the tanker. During this initial stage, crude oil is released to the ground and enters into the plant sewer (see Figure 6–2). In addition, the quantity of crude oil retained in the hose after unloading is released to the ground and therefore to the sewers (see Figure 6-3). The released oil results in dangerous slippery conditions for workers.

IER Case Study

Figure 6–1 Crude unloading station.

Figure 6–2 Releasing the remaining crude in the hoses to sewers.

Figure 6–3 The current unloading operation.

A field test was conducted to determine the unaccounted oil losses going into the sewer, which is transferred to be used later as fuel. The quantity of oil lost in the area was determined to be 10 tons/day as pure crude oil (tested at the Refinery Lab; Table 6–1 provides the oil separator analysis results). Taking into consideration that crude oil costs approximately $253/ton and is sold as fuel at a cost of approximately $112/ton, the unaccounted losses are $141/ton. This means that $1,410/day or $500,000/year are lost. This waste represents a continuous material loss, as thousands of dollars per year are sent to the sewer with only partial recovery of waste for use as fuel after separation.

The lost oil adds an enormous load to the on-site wastewater treatment plant (WWTP) and creates a major solid waste problem as sludge from the treatment plant. The solid waste (sludge) is managed with sludge/drying ponds. The sludge ponds are recognized as a significant long-term liability for the refinery.

Table 6–1 Crude Oil Unloading Station Oil Separator Analysis Results

		Sample Lab Results				
Number	Date	Time	Water Content (vol. %)	Sediment (vol. %)	API degree (at 60°F)	Specific Gravity (60/60°F)
1	3/30/05	8:00	0.0	0.8	31.71	0.867
2	3/30/05	10:00	17.5	0.6	31.89	0.866
3	3/30/05	13:00	0.0	1.0	31.89	0.866
4	3/31/05	9:00	0.0	0.8	31.52	0.868
From crude tanks			0.0	1.0	32.84	0.861

In addition, the crude overloads the American Petroleum Institute's (API) separator at the wastewater treatment plant. This reduces the treatment plant's efficiency and increases the quantities of sludge.

The following low-cost P2 recommendation was made and adopted by the refinery:

1. Instruct operators to use the existing hose carriers during and after unloading to prevent release of the residual crude to the sewer.

2. Provide caps for hoses and instruct the workers that they should be replaced after completing a tanker discharge. The caps are a low-cost investment ($7–$12/cap).

3. Prepare a written procedure for the unloading station and increase operator awareness of the cost associated with the losses and the negative impacts associated with current practice.

4. Recycle the uncontrolled releases of crude from the tankers back to the loading station by installing a collection tank and transfer pump at the API separator, as illustrated in Figure 6–4. The estimated cost is around $45,000 or less (including hopper tank, two explosion-proof pumps, fittings, and caps). The payback period based on a simple ROI (return on investment) calculation is about one month.

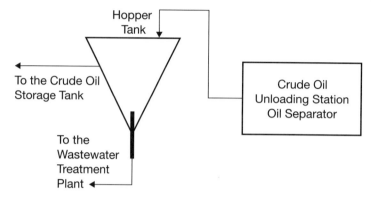

Figure 6-4 Proposed recycle tank configuration.

Liquid Petroleum Products Loading Station

Refined petroleum products are loaded to tankers using loading arms designed for this purpose. The loading station is divided into two sections, the first is used for light-end liquid products, such as gasoline, kerosene, and diesel, and the second is for heavy products, which include fuels and asphalt.

Tanker loading or charging is accomplished by filling the tanker from the top with vents that are open to the atmosphere. This results in displacement of vapors and creates significant fugitive emissions.

To eliminate the emissions the practice should be modified by changing the loading to a bottom feed configuration (see Figure 6–5)[2] and provide a vapor return line from the top of the vessel to either a vapor recovery unit or to the existing flare stack.[3]

An additional environmental aspect is spills. The existing loading arms are equipped with single-stage shutoff valves. A single-stage shutoff valve leads to frequent spills during charging from overfilling (see Figures 6–6 and 6–7). Spillage resulting from this

---

[2]Available at www.loadtec.co.uk/02-ptl.htm.
[3]Available at www.porttechnology.org/journals/ed12/pdfs/pt12-195.pdf and www.wodfieldsystemsindia.com/bottomloading-arms.htm.

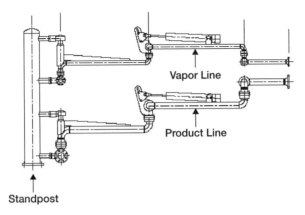

Figure 6–5 Bottom loading operation.

Figure 6–6 Petroleum products leaks.

Figure 6–7 Petroleum products leaks are drained into the sewer.

practice is drained into the sewer, causing additional loads to the WWTP. The accumulation of these spills represents losses in the products, which over time translate into significant financial losses. To minimize spills and avoid tanker overfilling, an automatic pressure control valve is recommended along with a two-stage shutoff valve to the loading arms.[4]

Loading LPG Bottling Plant

LPG cylinders are filled via semiautomatic equipment in the bottling plant. Liquid surfactant is used in the bottling plant for lubricating the conveyor during operation. Usually, the use of any surfactant leads to a reduction in the efficiency of physical oil treatment processes at the wastewater treatment plant, because it results in the formation of an emulsion that is difficult to break up. The emulsion

---

[4]Available at http://www.woodfieldsystemsindia.com/loading-arms.htm.

causes excessive foaming at later stages in treatment. The wastewater generated from the LPG plant is alkaline, with a very high TDS (pH = 10.8, TDS = 16,524 ppm, low FOG).[5]

A moderate-cost P2 recommendation aimed at improving the wastewater treatment plant efficiency is to stop sending the LPG bottling plant wastewater to the sewer. Instead, this stream should be diverted to the dissolved air flotation (DAF) unit, bypassing the API separator, which is the first stage at the wastewater treatment plant.

A second environmental aspect identified was that there is not enough ventilation in the plant. The entire area was congested with LPG, which is released during the semiautomatic filling process. As a mitigation measure, the refinery should install a ventilation system to reduce the health risks to workers.

Transport Section

The refinery maintains its own fleet of tankers for distributing petroleum products throughout the country. A separate maintenance and cleaning department supports the fleet.

During maintenance, lube oil from the tankers is discharged into the sewer (see Figures 6–8 and 6–9). The quantity of this spent oil is estimated to be 40 ton/year.[6]

This represents a significant amount of oil going into the sewer and hence adds to the oily and organic loads that shock the wastewater treatment plant. The additional cost to the WWTP associated with treating this is unknown, but clearly this action further reduces the API separator efficiency and the overall efficiency of the wastewater treatment plant.

The following no-cost P2 recommendation has been adopted to minimize the loads and help improve waste oil management practices in the workshop area:

1. Educate the workers on how to handle the spent lube oil.
   Operators should be using existing barrels to collect the spent

---

[5]From the Stone & Webster report.
[6]Engineer Talab Al-Jondy, manager of transport section.

Figure 6–8 The workshop area with poor housekeeping.

Figure 6–9 Discharging the lube oil into the sewer (workshop).

lube oil. This oil may be staged and recycled to the finished product fuel tanks. Since the spent fuel oil is sold at $113/ton, the estimated overall savings is $4,789/year.

2. Increase worker awareness about the downstream problems created by discharging this waste into the sewer and the recovered cost associated from the recycling.

3. Excess amounts of detergent are used in cleaning and washing operations. Since the detergents act as surfactants, they reduce the effectiveness of the physical oil treatment processes in the wastewater treatment plant. By providing laborers with a pressurized steam-cleaning machine, the amount of both detergents and water used in washing can be reduced by at least 30%.

4. Miscellaneous wastes, such as used vehicles parts, used batteries, oil filters, and tires, are haphazardly stockpiled (see Figures 6–10 through 6–12). This creates an unsafe work environment plus adds to site contamination from oils and cleaning solvents

Figure 6–10 The used vehicles are collected inside the section.

Figure 6–11 Trash heaps pile up at the facility.

Figure 6–12 Trash collection area, causing land pollution and long-term liabilities.

that drip onto the ground. It is recommended to allocate a special area where these wastes can be staged, segregated, and usable materials reclaimed.

Another environmental aspect is at the transport section, which has its own oil/water separator. The separated oil is sent to the API at the WWTP and the water is discharged into a ditch, leading off-site. A pool of contaminated wastewater exists next to the separator and frequent spillage to the ditch occurs (see Figures 6–13 and 6–14). This is a violation of environmental water discharge regulations and should be corrected. The P2 recommendation is to stop sending this oil to the API separator and, instead, collect it in the slop tanks, where it can be recycled or reused. This P2 recommendation needs to be assessed based on the quality of the oil collected. In the interim, the refinery should remediate the pool, stop discharging into the ditch, and send the waste stream to the wastewater treatment plant.

Technical Services and Laboratory

The laboratory section is equipped with instruments to perform all tests and analysis related to crude oil, petroleum products, lubricating oils, water analysis, and other analysis related to the operation, quality control, and environmental sampling.

Improper housekeeping practices were observed, especially in the collection of the empty chemical bottles and containers (see Figures 6–15 and 6–16). Hydrocarbon samples and hazardous and toxic chemicals are drained into the sewer. The following no-cost P2 recommendations have been adopted:

1. All chemical bottles, empty containers, and expired chemicals should be handled as a hazardous waste: collected, stored, and then sent to a secure landfill.

2. Collect the hydrocarbon samples in containers and recycle them to one of the oil slop tanks.

3. Educate the laboratory personnel on the safe handling and management of chemicals.

Figure 6–13 Oil/water pool near the oil separator.

Maintenance Sections

The maintenance sections are responsible for daily and preventive maintenance for all facilities. An environmental aspect observed is that the waste materials and trash generated during normal process unit maintenance practices are haphazardly stockpiled. Wastes and refuse consist of insulation materials (asbestos), oily sand, sludge, metal wastes, spent chemicals used for cleaning, and contaminated cleaning cloths (see Figures 6–17 through 6–19). This housekeeping problem has been eliminated by the following no-cost P2 recommendation:

IER Case Study 227

Figure 6–14   Oil/water spillage discharged into a ditch.

1. Segregate the leftover materials into recycle and waste materials. Store the useful parts in the store and dispose of the remaining materials in a secure landfill.

2. Develop written procedures for cleaning practices. These procedures should include proper waste management and disposal.

Figure 6–15 Empty chemical bottles pile up.

Figure 6–16 Empty chemical bottles reflect poor housekeeping at the lab.

IER Case Study 229

Figure 6–17 Debris and spent materials from maintenance operations.

Figure 6–18 More examples of debris and spent materials from maintenance operations.

Figure 6–19 Still more examples of debris and spent materials from maintenance operations.

Another environmental aspect is the leakage of steam, water, petroleum products, crude oil, chemicals, and firewater all around the refinery, indicative of an ineffective preventive maintenance program. The buildup of these leakages causes significant material loss, hence, significant financial losses. The following P2 recommendation has been adopted:

1. An aggressive preventive maintenance program with a parts tagging system.[7]

---

[7]Available at http://www.scafftag.com/products.asp?Index=3 and http://www.scafftag.com/products.asp?offset=10.

2. Implementation of a formal leak detection and repair (LDAR)[8] program.

3. Implementation of a computerized steam trap management system.[9]

Operation Department and Tank Farm

In Operations 1 and 2/tank farm section for crude oil, intermediate and finished products are stored in storage tanks. The staff are responsible for dosing chemical additives to the finished products.

The team observed a large number of spills. There is no spill prevention and response program. Spills are permitted to drain into the sewer, which adds to additional shock loadings to the wastewater treatment plant (see Figures 6–20 through 6–25). Spills occur because the dosing and tank filling operations are performed manually, often resulting in overfilling incidents.

To minimize spills, improve the safety of the work environment, and reduce pollution and loads to WWTP, the following low-cost P2 recommendation has been adopted:

1. Instruct operators to collect the spillage during the unloading operation and clean the area instead of using sand. Excess oil should be collected and recycled to the storage tanks.

2. Increase operator awareness of both the cost associated with the losses in this area and the negative environmental impacts.

3. Replace the existing cleaning method using water and detergent with a pressurized steam-cleaning method; this will minimize sliding on the roads[10] and reduce pollution.

---

[8]Available at http://www.pacndt.com/index.aspx?go=products&focus=Leak%20Detection.htm.
[9]Available at http://www.tlv.com/en/product/internat/fb5/fb5pdf/e-tm000-hp.pdf.
[10]Available at http://rea-na.com/MI-brochure.htm.

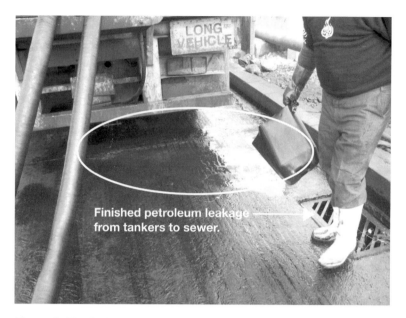

Figure 6–20 Spills and drainage from tanks to plant sewers cause downstream problems with the wastewater treatment plant.

Figure 6–21 Spills and unsafe conditions in the unloading area.

Figure 6–22 More examples of spills and unsafe conditions in the unloading area.

Figure 6–23 Other examples of spills and unsafe conditions in the unloading area.

Figure 6–24 Tank bottom (sludge) collected and stockpiled.

4. Implement a maintenance tagging system[11] program coupled with an LDAR[12] program and a computerized steam traps management system[13] program.

5. Install automatic liquid-level controllers to prevent tank overfilling.[14]

6. Chemical dosing is done manually: the operator carries the chemical (about 25 kg) to the top of the tank and adds the material, which is an unsafe practice. It should be stopped and addressed in the safety and health action plan. Workers need to be educated on the health and safety risks associated with

---

[11] Available at http://www.scafftag.com/products.asp?Index=3.

[12] Available at http://www.pacndt.com/index.aspx?go=products&focus=Leak%20Detection.htm.

[13] Available at http://www.tlv.com/en/product/internat/fb5/fb5pdf/e-tm000-hp.pdf.

[14] Available at http://www.oil-in-water.net/arjay_gateway.htm and http://www.simplexdirect.com/FuelSupply/controllers.html.

Figure 6–25 More tank bottom (sludge) collected and stockpiled.

such practices and an automated chemical dosing system installed[15] for this purpose.

7. The bottoms of crude oil storage tanks contain a large percentage of the sludge and pose a difficult disposal problem due to the presence of heavy metals. Tanks bottom sludge is composed of heavy metals, solids, water, rust, and scale (see Figures 6–26 and 6–27). Minimization of tank bottom sludge is carried out most effectively through careful separation of the oil and water remaining in the tank. When the oil storage tanks are

---

[15]Available at http://www.automatedaquariums.com/aasdose.htm.

Figure 6–26 Examples of steam and product losses and improper housekeeping.

drained for maintenance or other operational reasons, the accumulated sludge (slop wax) at the bottom of the tank also is drained into the plant sewer and the immobilized portions are collected in barrels and sent to the refinery's landfill. A mitigation measure that the refinery should adopt is an automated tank cleaning method and oil recovery system.[16] This system will help improve the oil recovery rate from the slop wax and decrease the loads to both the WWTP and the sludge ponds.

8. A considerable amount of water is frequently used for all pressure tests and leak check tests of the storage tanks. Once the test is finished, the used water is drained to the storm-water sewer network. This water can be recycled for cleaning purposes or used as firewater; however, to be able to use it, the

---

[16]Available at http://www.oreco.com/sw2800.asp.

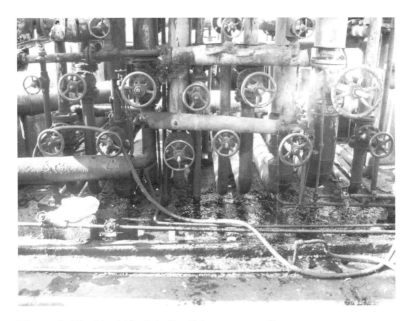

Figure 6–27 Liquid leak in the B-L vacuum unit.

tanks should be cleaned before the test to eliminate the presence of any impurities.

9. The tank farm sewers are directly connected to the WWTP to handle any accidental oil spills. These spillages shock the API separator and the WWTP. It is recommended that the refinery use one of the unused oil separators available close to the tank farm to recycle separated oil to the oil slop tank.
10. The refinery should prepare and implement a written spill prevention and contingency response plan.

Process Section

Crude oil is refined in the processes section to produce the final petroleum products. In these sections, leakages of steam, condensate, water, chemicals, crude oil, and petroleum products from pumps, valves, and steam traps were widely observed. These leaks are continuous and large in number, suggesting sizable material and product losses, and hence financial losses to the refinery. Figures 6–28 through 6–31 illustrate the environmental aspects. To reduce the losses and improve the overall performance of the refining section, the following P2 recommendations have been accepted:

1. Establish a well defined preventive maintenance program with combination with one of the following programs: maintenance tagging system,[17] LDAR,[18] a computerized steam traps management system,[19] and maintenance management system.[20]
2. Prepare and implement a spill prevention and contingency response plan to supplement the preventive maintenance program.
3. Implement a formal tracking system to account for losses and the incremental savings derived over time from these programs.

---

[17] Available at http://www.scafftag.com/products.asp?Index=3.
[18] Available at http://www.pacndt.com/index.aspx?go=products&focus=Leak%20Detection.htm.
[19] Available at http://www.tlv.com/en/product/internat/fb5/fb5pdf/e-tm000-hp.pdf.
[20] Available at http://www.scafftag.com/products.asp?offset=10.

IER Case Study 239

Figure 6–28 Steam and condensate leaks at vacuum stations.

Figure 6–29 Major steam losses from the steam traps (TOPING 3).

Figure 6-30 Leaking chemical pumps, reflective of a lack of an effective preventive maintenance program.

Sludge Ponds

The refinery has seven sludge ponds used for managing the waste generated from process operations (see Figures 6–32 through 6–35). These ponds are not lined and frequently overflow into nearby drainage ditches leading off-site.

The refinery has no clear plan on the elimination and remediation of these ponds; however, they pose a threat to the groundwater aquifer and neighboring community.

These ponds need to be remediated. They pose a long-term environmental threat and a likely public health threat. The P2 recommendations described earlier that reduce sludge generation in the wastewater treatment plant will reduce future sludge management issues for the refinery. However, the refinery still must address the current problems.

The following general recommendations are made:

Figure 6-31 Another leaking chemical pump's reflective of a lack of an effective preventive maintenance program.

1. The refinery should implement a study aimed at quantifying the amount of sludge contained in the ponds.
2. Groundwater monitoring wells should be installed to assess whether the aquifer has been impaired from permeation through the soil.
3. A review of cost-effective remediation technologies for both soil and contaminated groundwater should be made and recommendations made to senior management on the best course of action.

Figure 6–32 One of many unlined sludge ponds.

Figure 6–33 One of many unlined sludge ponds under construction on-site.

Figure 6–34 Sludge collected in barrels.

Figure 6–35 One of many toxic sludge ponds.

4. A series of lined lagoons should be constructed as an interim solution to the proper management of the sludge and for temporary staging of wastes from the current lagoons.

Spent Soda Drying Pond

Spent soda from the LPG recovery and chemical units were deposited in a drying pond for many years (see Figure 6–36). Recycling soda at the distillation process began about three years ago.

The drying pond is an environmental aspect that falls into the remediation category. Stressed vegetation can be observed around the pond, and there appear to be low-level gaseous emissions especially in wintertime.

The aspect merits a closer examination on the part of the refinery. The refinery should reevaluate security access to the pond to ensure animals, the public, and workers are not exposed (i.e., a

Figure 6–36 This spent soda drying pond will exist for many years, since it is a stable emulsion.

fence should be constructed around the perimeter of the pond). It should also check the integrity of the lining with a groundwater monitoring well and assess options for remediation.

Sour Water Strippers

Sour water from the refining section is sent to the sour water striping system to remove H_2S, NH_3, and phenols. There are two sour water strippers.

Stripper tower no. 1 processes approximately 42 m³/h of sour water from CDU-1, CDU-2, FCC, and vacuum no. 1 unit. The operation is done by injecting steam in the bottom of the tower. Off-gasses (H_2S and NH_3) are discharged directly to the atmosphere, and the stripped water is directed to stripper no 2.

Stripper tower no. 2 (Figure 6–37) processes around 16 m³/h of sour water from CDU-3, NHT, UNIBON, NHT, and stripper no 1. The produced gasses are incinerated (Figure 6–38), and the stripped water is discharged to the WWTP, bypassing the API separator.

The major environmental aspect is the air emission problem (H_2S). The high concentration of H_2S and NH_3 in the generated wastewater from the strippers[21] indicates low stripper efficiency, which also needs further investigation.

Hydrogen sulfide (H_2S) is a colorless gas with an offensive stench and is said to smell like rotten eggs. The gas can be detected at a level of two parts per billion. Hydrogen sulfide has few important commercial uses. However, it is used to produce sulfur, which is one of the most commercially important elements. About 25% of all sulfur is obtained from natural gas and crude oil by conversion of one-third of the H_2S to SO_2, followed by the reaction between H_2S and SO_2:

$$2\ H_2S(g) + 3\ O_2(g) \rightarrow 2\ SO_2(g) + 2\ H_2O(g)$$
$$16\ H_2S(g) + 8\ SO_2(g) \rightarrow 3\ S_8(g) + 16\ H_2O(g)$$

The refinery should consider an investment in a sulfur recovery unit, a high-cost P2 investment. Approximate costs for this investment

---

[21] Refinery wastewater analysis.

Figure 6–37 Stripper no. 2.

are in excess of $7 million. A separate study should be commissioned to evaluate the investment and assess cost recovery options. Jordan has a strong fertilizer manufacturing market, which would take sulfur at market rates.

Tetraethyl Lead Station (Overall Health and Safety Issues)

Tetraethyl lead is added to gasoline to increase the octane number. The first environmental aspect is a significant worker safety risk. Workers were observed not wearing respirators and helmets during the operation, and smoking, drinking, and eating in the control room was observed during the walkthrough. Work uniforms are

IER Case Study 247

Figure 6–38 The stripper no. 2 incinerator.

not properly decontaminated. The refinery should implement a comprehensive health and safety plan that encompasses training on safe work practices and chemical handling.

A second environmental aspect relates to filled barrels of tetraethyl lead, which are collected and haphazardly staged and discarded about the station. The empty barrels are burned periodically, adding to the already stressed air pollution emissions from the facility. These can and should be recycled to the supplier (see Figure 6–39).

A third environmental aspect is that the staging area for spent barrels also is used as a landfill. Contaminated chemical protective clothing was observed partially buried (Figure 6–40). Since these

Figure 6–39 Spent tetraethyl lead barrels are permitted to stockpile on-site, a poor environmental practice.

materials cannot be decontaminated, they become a part of the hazardous waste stream from the operation and must be handled and disposed of in a proper manner.

The following no-cost P2 recommendations have been adopted:[22]

1. Provide safety equipment needed for this section, especially chemical protective clothing and respirators.
2. Implement a respiratory-fit testing program.
3. Establish staging, decontamination, and disposal procedures for contaminated clothing and respirators.

---

[22]*Guidance Document for Design of Worker Safety Training Program for the Jordan Petroleum Refinery Co.*, February 2005, CDM International Inc.

Figure 6–40 Contaminated work clothes in an on-site landfill.

4. Educate and train workers on health and safety risks that are job specific and on the safe handling of hazardous and toxic chemicals.
5. Return the empty barrels to the manufacturer and dispose of the contaminated chemical protective clothes at a secured landfill designed for hazardous wastes.

There is a significant lack of protocol and practices aimed at addressing health and safety throughout the refinery. For examples, safety equipment are not used during crude unloading operations; in the transport section, workers were not wearing PPE (personal protective equipment) like safety shoes, respirators, helmets, and chemical protective clothes. Also in the operation 1 and 2 tank farm, there is a general lack of safety awareness and security control. In addition, improper electrical grounding was observed in a number of operations.

Senior management needs to understand that worker safety-related aspects are a concern not only from a lost productivity standpoint but have a direct negative impact on insurance premiums and ratings. Integration of the safety department within the EMS would reduce risks and financial exposures significantly.

General No-Cost/Low-Cost P2 Recommendations

Good housekeeping practices will eliminate wastes from piling up, reduce worker accidents, minimize on-site contamination, reduce the risks for fires, and improve overall productivity. Senior management has agreed with this and initiated adoption of the following recommendations.

Industrial Solid Wastes

A large amount of nonhazardous solid waste is collected in a boneyard that is located next to the flare. Further assessment showed that both hazardous and nonhazardous wastes are mixed together (see Figures 6–41 and 6–42).

It is recommended that theses solids be segregated as hazardous and nonhazardous solids. Hazardous wastes can be disposed of in secure landfills designed for such wastes, and the nonhazardous stream may be deposited in a municipal landfill.

General Housekeeping

In the processes sections, such as the asphalt, distillation, and FCC units, and the lube oil, barrels of spent and waste materials are haphazardly staged and stockpiled. This results in oil and product spills (see Figure 6–43).

In the caustic soda solution unit, the caustic soda is placed in rusty and corroded barrels, leading to dispersion of the soda all around the unit (see Figures 6–44 and 6–45). Soda increases both wastewater salinity and metal corrosion. Used chemical PVC and steel drums are also collected and scattered about (see Figures 6–46 and 6–47).

IER Case Study

Figure 6–41 Neglected and mounting trash heaps add to land and possible storm-water runoff pollution.

Figure 6–42 More neglected and mounting trash heaps.

Figure 6-43 A stockpile of lube oil and chemical barrels, indicative of poor housekeeping.

Figure 6-44 Poor housekeeping during the preparation of caustic soda solutions.

IER Case Study

Figure 6-45 More poor housekeeping and unsafe practices during the preparation of caustic soda solutions.

Figure 6-46 Firewater leaks in the tank farm.

Figure 6-47 Chemicals drums piled up the tank farm is poor housekeeping.

Poor preventive maintenance was seen in the tank farm area. Steam, water, petroleum products, chemicals, and firewater leakages from valves, pumps, and steam traps were noticed everywhere (see Figure 6–48).

The following low-cost/no-cost P2 recommendations have been adopted by the refinery:

- Develop a written housekeeping plan and procedures for each section of the refinery.
- Use dry cleaning as a first option in case of any oil or chemical spillage and replace the existing washing techniques using a flow of water with a pressurized steam-cleaning method to improve cleaning efficiency and reduce water consumption and, therefore, wastewater generation.[23]
- Store lube oil barrels in organized stores, well labeled and defined.
- Place caustic substances in plastic bags. Construct a store area for caustic soda bags to prevent any soda losses.

---

[23] Available at http://rea-na.com/MI-brochure.htm.

Figure 6–48 A collection of filled gasoline barrels in the tank farm.

- Send spent chemical PVC and steel drums to the allocated staging area for disposal.
- Implement a waste inventory and tracking system, so large amounts of wastes do not continue to stockpile on-site.

Service Water Tank: Mountain Tank

The services tank, or the mountain tank, is a concrete tank located in the highest part of the property (to allow gravity flow) and provides the refinery with the required service water pumped from wells. The team noticed water leakage from cracks in the tank. Since the level controller is not working effectively, the refinery should repair it and make necessary repairs to the vessel to stop the leakage.

Firefighting Network

The team also observed leakages from the firewater network all over the refinery. This needs to be maintained and considered in the regular preventive maintenance program.

Figure 6–49 Slop tanks.

Slop Tank

The refinery has two groups of slop tanks. The first group is used to recover oil from the API separator, and the second group is located in the tank farm (see Figure 6–49). The oil separation from wastewater in these tanks depends on the physical process. The contents of the tanks are heated to 66°C over a period of three hours. Once the desired temperature is reached the separation process is completed, the bottom layer, which contains the wastewater, is drained to sewers through a valve located at the lowest point of the tank and the top layer, which contains the oil, is pumped to the final product fuel tanks (slop tanks, second group).

During the IER, most of the sewers near the slop tanks (the second group) were observed to be partially blocked. The area became a pool of oily water (see Figure 6–50). This condition results from a poorly implemented preventive maintenance program. Substituting pressurized steam-cleaning methods for existing washing techniques

IER Case Study

Figure 6–50 Oil/water pool in the slop tanks area.

will improve both housekeeping and water conservation.[24] Sewer lines, sumps, and catch basins also need to be cleaned periodically.

Another environmental aspect observed regards draining operations. The draining operation of the bottom layer in the tanks, which contains wastewater, is done manually, which often leads to spills. Large slugs of oil were observed to spill directly onto the ground and drain into the sewer. This creates shock loads to the API separator. The P2 recommendation is to replace the drain valves with higher-density sensitive valves (i.e., valves with a sensor that detects density or conductivity differences between oil and water and automatically closes if oil is detected).[25] Additionally, this valve may be used for the following purposes:

---

[24]Available at http://rea-na.com/MI-brochure.htm.
[25]Available at http://www.cobhamfluidsystems.com/products_control_valves.htm#7.

1. On oil storage tanks. When draining water from the bottom of tanks, they will prevent fuel or oil carryover and minimize the risk of pollution.
2. Installed on the rainwater drain line, they will automatically close if oil or fuel is detected in the water stream.

Another environmental aspect concerns oil analysis. Oil analysis (water content, sediment) is performed on the separated oil in the slop tanks (second group) before sending it to the fuel product tank. Oil that does not meet the standards is drained into the sewers. This plugs the sewers and increases the housekeeping problems in the area. The pumping cost (from the slop tank, first group, to the second group) and the separation cost in both groups represent a financial loss.

To overcome this problem the refinery should change the oil analyzing program by analyzing the separated oil in the slop tanks in the first group (once the separation operation is finished). This P2 recommendation will enable the refinery to save both money and time and increase the equalization of the received wastewater at the API separator.

Final Remarks

In this case study more than $1.4 million in P2 savings have been quantified and 250,000 m^3 of water savings identified in the utilities section of the refinery. Based on similar exercises conducted at refinery operations in Aruba, Romania, and the United States, the facilitywide IER conservatively identified another potential savings on the order of $1 million in materials, energy, productivity, workhours, and reduced insurance premiums.[26] All these savings were identified from the IER in the early stages of the EMS implementation.

The case study is a good example of how elements of an EMS can improve both the financial and environmental performance for

---

[26]This is from my experience as lead practitioner in resource reduction and P2 audits at the Exxon Aruba refinery, the Exxon Bayway New Jersey refinery, and the Rompetrol Refinery in Romania.

a company. While the application of an EMS traditionally focuses on compliance, particularly in the United States, a major benefit to any company is that it enhances the competitiveness and sustainability of the enterprise.

While many of the baseline environmental practices identified in the exercise would never be observed in U.S. refinery operations today, due to strict environmental enforcement of RCRA (Resource Conservation and Recovery Act) and other key environmental legislation, the case study does emphasize that significant environmental improvements and savings are possible, even at the early stages of an EMS implementation. Some skeptical readers might argue that the savings and financial gains for this facility were relatively easy to identify and capture, since the absence of strict enforcement in Jordan enables even simple housekeeping improvements to make a significant positive difference. To those skeptics, I note the following. While U.S. domestic operations are always on guard against the pending liabilities from strict environmental regulations and enforcement, in many countries like Jordan, where the threat of environmental enforcement is low and there is even a lack of legislation, business leaders are even more skeptical of the benefits of an environmental management system. The mentality among business leaders in such countries is that pollution management is a part of the cost of doing business, and it is given very low priority. But, when financial gains are coupled with good environmental performance, the value of an EMS suddenly grabs their attention and support.

7
Cleaner Production Case Studies

Introduction

This chapter explores the power of cleaner production (CP) and pollution prevention (P2) to refinery operations. CP and P2 programs work best within the context of an EMS. In fact, ISO 14001 emphasizes that P2 is the core of the environmental management system.

The case studies included in this chapter are but a random sampling of what is reported by companies. Most have been taken from open literature sources, while others are ones where I have been either in communication with corporate representatives or had some personal involvement through project assignments. The examples are from a range of companies that represent both major and smaller players in world markets. A key feature is that they are international in scope, which is a testament to the fact that all companies face the same fundamental issues—sustainability and competitiveness.

As noted in earlier discussions, no substantive distinctions are made between the terms *cleaner production* and *pollution prevention*. While the latter generally implies that the focus of efforts is on reducing pollution, often of a regulatory nature, all forms of waste and, in a more general sense, inefficiency may be thought of as forms of pollution. In a perfect world, 100% of the raw materials, energy, and productivity that goes into a manufacturing process would result in the transformation of useful products that serve the needs of society. But engineering and manufacturing have limitations that produce losses as wasted energy, harmful by-products, and emissions that can result in contamination of the environment and even place the public at large at risk from exposures during the creation of useful products. Setting aside environmental damages and human health risk considerations, companies are in business to make a profit, and most companies have the vision to continue to meet the demands of future generations. Consequently, a core value of any company that acts responsibly, whether it be from the standpoint of protecting the environment and public health or from sound business principles, is to make itself as efficient as possible. Herein lies the fundamental reason why cleaner production and P2 should be the cornerstones of any refinery and, in fact, any business.

CP/P2 Case Study at the Jordan Petroleum Refinery Co., Zarqa, Jordan

Applying the 21-step procedure for conducting a CP/P2 audit described in Chapter 3, the refinery's pollution prevention team implemented an audit of the utilities section of the refinery. Background information on the refinery along with a case study of an IER conducted at the facility can be found in Chapter 6. The financial savings reported for various opportunities are reported in U.S. dollars ($). A currency equivalent of $0.71 per JD (Jordanian dinar) was used.

CP/P2 Opportunities for Boilers and Steam Recovery

The refinery has six water tube boilers with a combined steam production capacity of 130 ton/h. The boilers produce three types of steam (low-, medium-, and high-pressure steam) and generate two wastewater streams. By applying material and energy balances the team identified significant losses attributed to boiler blowdown and excess condensate.

As water evaporates in the boiler steam drum, solids or scale present in the feedwater are left behind. The blowdown purpose is to limit the concentration of impurities in the boiler water, which can cause scale formation and corrosion of tube walls, thereby reducing heat exchange efficiency as well as damaging the boiler (Table 7–1 shows the maximum limitations for boiler water and feedwater). The right amount of blowdown is critical. Too much can result in energy losses and excessive chemical treatment cost; while too little will allow excessive concentrations of impurities to build up and damage the tubes.

In water tube or drum-type boilers, there are two blowdown systems: continuous blowdown from the steam drum to limit the concentration of the water and intermittent blowing from the bottom headers to remove suspended solids from the boiler.

During the assessment, the total boiler blowdown rate was found to be 6m^3/h. This entire amount is drained to the refinery's wastewater treatment plant. Minimizing this blowdown rate reduces

Table 7–1 Limitations on Boiler Water and Feedwater

Item	Feedwater	Boiler Water
pH (at 25°C)	7.5–9.0	10.0–11.0
T. Hardness as CaCO$_3$	1	—
Iron (ppm)	0.05	—
D. Oxygen (ppm)	0.05	—
Copper (ppm)	0.05	—
TDS (ppm)	12.5	500
Conductivity (μs/cm^2)	30	1200
Silica (ppm)	0.5	20
Phosphates (ppm)	—	15–25
Sulfates (ppm)	—	10
Oil keep	0.0	—
P-alkalinity as CaCO$_3$	—	80–300
Chloride (ppm)	—	35*

*From the assistant operation engineer for area 4 (utilities).

materials and energy losses. Since the temperature of the blowndown water is the same as that of the steam generated in the boiler, any savings in blowdown will also reduce the makeup feedwater and the chemical treatment additives.

Tables 7–2 and 7–3 report the water analysis for the boiler blowdown. The instability in the controlling process is obvious from the data, indicating that the blowdown system is ineffective. Ineffective control, reflected by the data in Table 7–1, leads to excessive blowdown. Figures 7–1 through 7–4 summarize the issues.

The silica concentration was used to calculate the boiler's concentration ratio, since it is a critical controlling limit for water tube boilers. Silica comes from several types of deposits that are considered very difficult to remove (e.g., amorphous silica[1]).

Table 7–2 Boiler 7001 Blowdown Analysis

Sample Date	pH	Conductivity ($\mu s/cm^2$)	Chlorides as $CaCO_3$ (ppm)	Phosphate content (ppm)	Hydrazine (ppm)
1/3/05	11.3	253	56	2	0.05
1/10/05	10.6	192	56	2	0.05
1/17/05	10.4	529	36	0	0.2
1/24/05	11.7	316	39.2	2	0.2
1/31/05	11.5	496	28	0	0.05
2/7/05	12	922	22	4	0
2/14/05	11.6	525	42	0	0
2/21/05	10.8	540	17	3	0.4
2/28/05	10.8	412	17	1	1
3/7/05	10.5	430	31	2	0
3/14/05	10.2	210	17	1	0.1

[1] F. N. Kemmer, ed., *NALCO Water Handbook*, 2nd ed. (New York: McGraw-Hill, 1988).

Cleaner Production Case Studies

Table 7-3 Boiler 7004 Blowdown Analysis

Sample Date	pH	Conductivity (µs/cm²)	Chlorides as CaCO₃ (ppm)	Phosphate content (ppm)	Hydrazine (ppm)
1/3/05	11.8	78	16.8	0	0.3
1/10/05	10.6	190	16.8	1	0.05
1/17/05	10.5	147	11.2	0	0.1
1/24/05	11.2	112	5.6	1	0.1
1/31/05	11.8	81	14	0	0.05
2/7/05	11.5	111	11.2	2	0.2
2/14/05	11.8	508	28	0	0
2/21/05	11.2	894	28	3	0.4
2/28/05	11.1	902	28	1	0.2
3/7/05	10.3	370	28	2	0

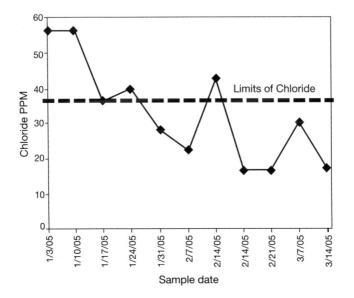

Figure 7-1 Boiler no. 7001 blowdown chloride control.

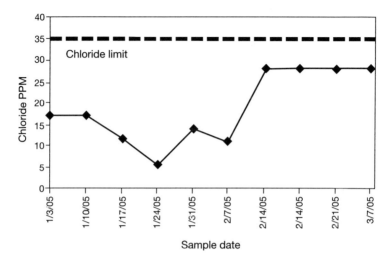

Figure 7–2 Boiler blowdown chloride control.

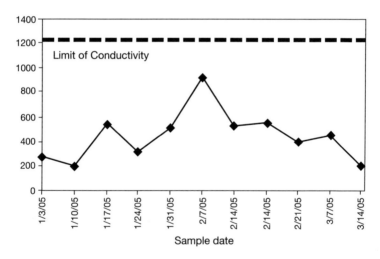

Figure 7–3 Boiler no. 7001 blowdown conductivity control.

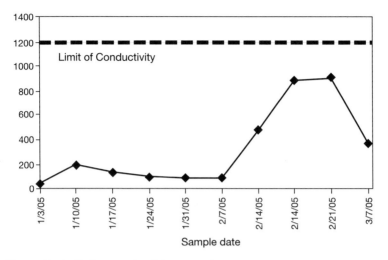

Figure 7-4 Boiler conductivity control.

The theoretical concentration ratio CR is

Theoretical CR = C_B/C_F
= 20/0.5 = 40

Theoretical B = F/CR
= 130 m^3/h/40 = 3.25 m^3/h

Actual CR = F/B
= 130 m^3/h/6 m^3/h = 21.7

Excessive blowdown = actual blowdown − theoretical blowdown
= 6 − 3.25 = 2.75 m^3/h

where

CR = concentration ratio;
C_B = the concentration of X mg/L solid in boiler blowdown water;
C_F = the concentration of X mg/L solid in the boiler feed water;
B = boiler blowdown; and
F = boiler feed water.

Note: The blowdown cost is influenced by the cost for high-pressure steam (around $10/m^3).[2]

$$\text{Excessive blowdown cost} = 2.75 \text{ m}^3/\text{h} \times \$10 \times 24 \times 365$$
$$= \$240{,}900/\text{yr}$$

The $240,900 is the total savings that could be achieved yearly by better control for the boiler blowdown and improved management of the operation.

CP/P2 Actions

To capture the savings, the refinery should change the current blowdown approach and optimize the boiler's operating window as opposed to performing random blowdown. In practical terms, this means automation.

To change from a manual system to automated control, the system illustrated in Figure 7–5 should be installed. This will eliminate operator error and provide system stability. An automatic blowdown control system[3] optimizes surface blowdown by regulating the water volume discharged in relation to amount of dissolved solids present, conductivity, and silica or chloride concentrations. A probe provides feedback to a controller/driving mechanism through a modulating blowdown valve. An alternative control configuration is a proportional control with the blowdown rate set proportional to makeup water flow.

Figure 7–6 conceptually illustrates the degree of controls between manual and automated control schemes for this application.

The approximate cost for an automated systems is $23,239. The total cost for six boilers is $140,844 with a payback period of seven months maximum.

---

[2] From the refinery team.
[3] Available at http://www.spiraxsarco.com/assets/uploads/PDFs/sb/P403_03.PDF.

Cleaner Production Case Studies

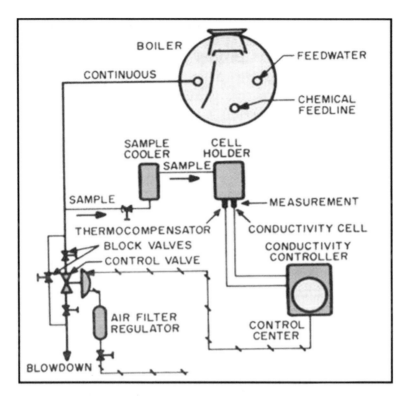

Figure 7-5 Automatic boiler blowdown system.

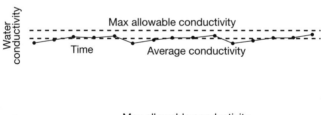

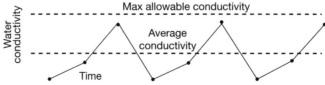

Figure 7-6 Manual (top) versus automatic (bottom) blowdown.

CP/P2 Opportunities for the Condensate Recovery System

When condensate leaves the steam trap, it has approximately 20% of the original thermal energy contained within the steam. Returning this condensate for reuse further reduces energy consumption, eliminates the excess use of water treatment chemicals, and allows a high quality of water to be reused as water feed makeup.

During the P2 audit the team observed a significant amount of steam and condensate losses. The root causes for these losses are that steam traps malfunction and steam is misused in washing and cleaning operations. The steam and condensate losses were difficult to estimate; however, they represent a sizeable loss that translates into thousands of dollars. Figures 7–7 though 7–9 provide examples of steam losses and misuse.

Figure 7–7 Unrecovered steam.

Cleaner Production Case Studies

Figure 7–8 Condensate is drained into sewers because of high chloride.

Figure 7–9 Steam is used for cleaning.

CP/P2 Actions

The following actions were recommended:

- Stop or limit inefficient steam cleaning and washing operations.
- Increase operator awareness of condensate management and losses.
- Repair steam traps to improve condensate recovery and overall steam system performance.

CP/P2 Opportunities for the Low-Pressure Condensate Tank

Because of malfunction of the level controller of the low-pressure condensate tank (see Figure 7–10), low-pressure condensate return cannot be controlled. This results in frequent condensate overflow. The overflow drains to sewers after being cooled with running service water (from well no .7) to minimize steam emissions in the working area (see Figure 7–11). This represents a significant water loss.

Figure 7–10 Condensate recovery tank.

Cleaner Production Case Studies

Figure 7–11 Condensate overflow is drained into sewers.

Draining the condensate to sewers is not only a misuse of valuable water, it adds to the energy deficit identified.

The team was not able to measure the amount of condensate overflow; however, the estimated minimum amount of service water used to cool this condensate is about 1 m^3/h. This represents 24 m^3/day or 8760 m^3/year of water savings. In terms of costs to the refinery, this is $3085/year.

CP/P2 Actions

To minimize the condensate and steam losses, the following actions were recommended:

1. Establish a preventive maintenance program to repair all steam traps. This will reduce the amount of condensate return and, therefore, the condensate overflow.
2. Repair the low-pressure condensate tank level controller.
3. Install a control valve at the service water line cascade along with a high-level indicator on the tank. This will minimize the service water losses.

CP/P2 Opportunities for the Intermediate Condensate Collecting Tank

During the P2 audit, a significant amount of flash steam was noticed open to atmosphere from the intermediate condensate collecting tank (see Figure 7–12). This is because the tank is uncovered and open for continuous natural evaporation. This observation also supports the conclusion of the failure of most steam traps in the operation sections; therefore, unusual excess amounts of steam are carried with the return condensate.

CP/P2 Actions

To prevent the loss of steam to the atmosphere and reduce the amount of steam carried with the return condensate, it was recommended to

- Cover the condensate tank and provide vents for the normal expected steam return.

Figure 7–12 Intermediate condensate tank.

- Implement a preventive maintenance program, as recommended earlier for all steam traps. This will help reduce the steam return with the condensate.

CP/P2 Opportunities for Steam Losses

Steam losses are caused by faulty steam traps and poor maintenance practices. Additional examples are shown in Figures 7–13 through 7–17. Minimizing steam losses will bring large savings in well water and fuel use, decrease the generated wastewater from the reverse osmosis (RO) and ion exchanger units, and indirectly reduce the chemical usage in RO, ion exchangers, and boilers.

Since there are no steam flow meters, the team was not able to calculate the exact amount of losses; however, a reasonable estimation was made based on the following:

- First estimation: Steam leakage reported as 15 m3/h.[4]

Figure 7–13 Steam leak in the boilers section.

---

[4]Personal communication with the assistant operation manager area 4 (utilities).

276 Environmental Management Systems Handbook for Refineries

Figure 7-14 Steam leak in vacuum 2.

Figure 7-15 Another example of excess steam leakage in the utility section.

Figure 7-16 Another example of excess steam losses associated with vacuum units.

- Second estimation: Steam leakage calculated from the refinery water distribution diagram is approximately 17 m^3/h.
- Third estimation: From a Stone & Webster unpublished report for the refinery, losses were estimated to be approximately 12 m^3/h.

The main steam loss as reported from the refinery team was from the medium- and low-pressure steam.

Taking into consideration the lower estimated steam losses, the calculated value of these losses is:

Total steam leakage amount/year = 12 × 24 × 365
= 105,120 m^3/yr

Total steam leakage cost/year = 105,120 × 6.00
= $631,608/year.

This is worth capturing!

Figure 7-17 Steam leak in TOP 3.

CP/P2 Actions

It is obvious that a preventive maintenance program is needed. Due to the large number of point sources, the best approach to implementing a preventive maintenance program is to invest in a computerized steam trap management system.[5] Figure 7–18 shows a commercially available unit that will help improve the overall maintenance program. The unit comprises a measurement and analysis system for diagnosing steam trap operations and quantifying steam leakage. PC-based software facilitates implementation of a comprehensive steam trap management system. In addition, the system provides a database for each steam trap.

[5]Available at http://www.tlv.com/en/product/internat/fb5/fb5pdf/e-tm000-hp.pdf.

Figure 7–18 Computerized steam traps management system.

The cost for such a system was found to be approximately $19,718. With the $631,608/year savings just calculated, the payback period is between 0.5 and 1 month.

CP/P2 Opportunities for the Reverse Osmosis Unit

The existing reverse osmosis (RO) unit (see Figure 7–19) is fed from well water. The unit produces treated water with conductivity 200–300 μs/cm^2 and generates reject water (brine) with conductivity 9,000–10,000 μs/cm^2. The water from the RO unit is makeup for boilers.

This unit was under replacement with a new and larger-capacity RO unit that would nearly change the overall water consumption data and material balance in the utility section. The team was unable to predict the expected changes but, for example, feeding the cooling tower with RO water would help reduce the blowdown and increase the concentration ratio for the operation.

The lower content of silica, chloride, and conductivity in the new RO unit water would help reduce the frequency for ion exchange regenerations and, therefore, decrease the generated

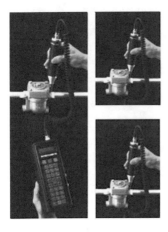 

Figure 7-19   Reverse osmosis unit.

wastewater, reduce water consumption, and reduce treatment chemicals use.

In addition, the use of this water as water makeup for the boilers would help reducing boiler blowdown and increase the concentration ratio of the operation, thus saving more water and chemicals.

Table 7–4 provides a comparison between the old and new units.

CP/P2 Actions for Flushing the RO Membranes and Sand Filter Backwash

We first focus on RO flushing. The current RO unit is flushed daily for one hour. During the audit, the team performed a simple analysis (Table 7–5 provides the analysis results) on the reject water dur-

Table 7–4   Old versus New Reverse Osmosis Unit

Test	Old RO	New RO
Maximum permeate flow (m^3/h)	100	180
Recovery	75–80%	80%
Conductivity ($\mu s/cm^2$)	329	35
Chlorides (ppm)	20	6.8
Silica as SiO_2 (ppm)	0.618	0.14

Table 7-5 RO Reject Analysis during Flushing (sample date: 4/4/2005)

Skid No.	TDS (ppm)	Flow for 1 Hour (m³/d)	Flow for 50 Minutes (m³/d)
A	3160	15.6	13
B	4112	9.6	8
C	3450	13.2	11
D	3450	11.4	9.5
Average	3543	Total amount	41.5

ing the flushing process and found that the first 5–10 minutes of the flushing results in the highest amount of salinity (high TDS), while the next 45 minutes of flushing produces brine that is in the range of 3500 ppm. This latter water quality is close enough to the refinery well water TDS (well no. 7 has a TDS of 2629) and therefore it is recommended to drain the first 10 minutes reject from the flushing and collect the rest. The estimated amount of this water is approximately 41.5 m³/day. This water can be reused for cleaning purposes or held as firewater.

The second P2 recommendation focuses on the backwash from the sand filters. The refinery has three sand filters (see Figure 7–20) as a pretreatment stage for the RO unit, where they help reduce the total suspended solids in the RO feedwater. Daily backwash is performed with about 20 m³ in 10 minutes. The total backwash water for the three filters is 60 m³/day. This water is currently drained into the sewer, although it could be reused for cleaning or as firewater since the TDS of this water is lower (1500 ppm) than water supply from the well (well water TDS is about 2629 ppm). Table 7–6 reports the sand filters backwash analysis results.

Savings from Reusing RO Flush Water and Sand Filters Backwash

The sand filter units are shown in Figure 7–21. The total amount of generated water, which later is considered part of the wastewater from the RO flushing and sand filters backwash, is 37,047m³/year: therefore, $13,044/year can be saved by reusing it.

Figure 7–20 Sand filters.

Table 7–6 Sand Filters Backwash Analysis (sample date: 4/4/2005)

Sample No.	Time (min)	Lab Tests		
		pH	TDS	TSS
1	0.5	7.6	1520	102
2	5	7.7	1480	100
3	9	7.6	1500	97

To collect this water, a holding tank with an approximate capacity of 120 m^3 is needed. This will cost \$21,127. With this estimation, the payback period will be in less than 2 years.

CP/P2 Opportunities for the Cooling Towers

The refinery has three cooling towers that consume approximately 120 m^3/h as a makeup water from wells (conductivity around 4,369 μs/cm^2). Figure 7–22 shows the cooling tower. About 45–50% of the cooling water is lost through evaporation, and the remaining

Cleaner Production Case Studies 283

Figure 7-21 Sand filter units.

Figure 7-22 TSEP cooling tower.

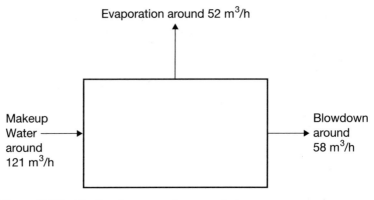

Figure 7–23 Cooling towers water mass balance.

percentage is discharged as blowdown with a conductivity of 5,000–6,100 µs/cm².

Figure 7–23 presents the cooling towers water mass balance. The high TDS content in the makeup water decreases the concentration ratio (around 1.7), increases water and chemicals consumption, and therefore blowdown frequency.

CP/P2 Actions

Since the refinery is installing a new RO unit and will be feeding the boilers and cooling towers with a higher quality of water, the overall performance of the cooling towers will increase. The water and chemical treatment consumptions will decrease and the total blowdown will decrease since the concentration ratio will go up with such high quality (Figure 7–24 demonstrates the expected improvement).

CP/P2 Opportunities for the Ion Exchanger Units

The refinery has two ion exchangers units that demineralize water that serves as feed to the boilers. Cycle time for both units is 4 to 5 days and NaOH and HCl are used for the resin regeneration. The generated wastewater from the regeneration process is collected in a neutralizing pit and used for neutralizing the wastewater at the

Cleaner Production Case Studies

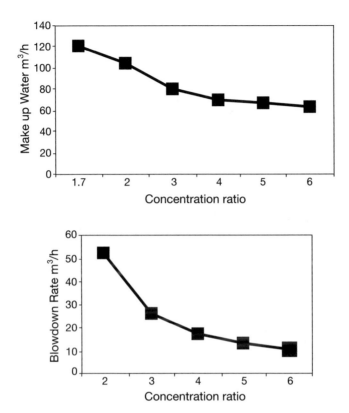

Figure 7-24 Present relation between concentration ratio, makeup water, and blowdown.

wastewater treatment plant, bypassing the API separator, since it has a pH of 2–3.

Since the total volume of the generated water from the regeneration is larger than volume of the neutralizing pit (see Figure 7–25), the backwash and a portion of the rinse water is sent to a pool near the cooling towers through a channel (see Figure 7–26). The pool is used to collect the sand filters backwash, part of the regeneration effluent, and sometimes part of the cooling tower blowdown. The pool water is diverted also to the API separator in the WWTP, which increases the amount of sludge generated from the plant to the sludge ponds.

Figure 7–25 Neutralizing pit.

CP/P2 Actions

Since part of the regeneration effluent has good quality, it can be reused. The generated water from backwash and rinsing has low conductivity; therefore, it could be used as cooling towers' makeup water, in addition to the RO water. However, the first 10 minutes of the generated water from the rinsing step have high conductivity, so it should be collected in the neutralizing pit with the rest of the regeneration water (regenerate, injection, and displacement stages).

The low-salinity part of the regeneration effluent (backwash and rinsing without the first 10 minutes) should be collected only in the southern pool. Since the volume of this pool is small (around 110 m^3) the cooling towers should be fed with part of this water during the regeneration operation (the duration of regeneration process is around 6 hours). Figure 7–27 shows the recommended reuse and disposal of the ion exchangers' regeneration effluent.

Figure 7–26 Channel sending filter backwash and regeneration effluent to a central pool.

Table 7-7 provides the volumes of the neutralizing pit, southern pool, and the other related quantities of regeneration process. Note that the sand filter backwashing water could be drained with the same RO reject pipe.

From Tables 7–8 and 7–9, the total quantity of the regeneration effluent from unit 1 is 301.9 m^3/cycle. The refinery could reuse 157.7 m^3. Also the total quantity of the regeneration effluent from unit 2 is 369 m^3/cycle. We calculate that the refinery could reuse 216 m^3.

288 Environmental Management Systems Handbook for Refineries

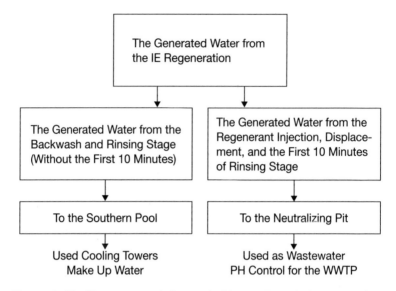

Figure 7–27 The reuse and disposal of ion exchangers' regeneration effluent.

Table 7–7 The Volumes Related to Ion Exchanger No. 1 and No. 2 Units

Items	Volume (m^3)
Neutralizing pit	200
Southern pool	110
Total quantity of no. 1 regeneration water	302
Total quantity of no. 2 regeneration water	369
Quantity of backwash and rinsing (without the first 10 minutes) water for no. 1	157.7
Quantity of backwash and rinsing (without the first 10 minutes) water for no. 2	216

Table 7-8 Ion Exchanger No. 1 Regeneration Analysis (Brief)

Unit	Sample	Lab Tests (sampled on 5/3/2005)				Total Quantity (m^3)	Backwash/ Rinsing Water (m^3)	Disposal Water (m^3)
		pH	Conductivity	TSS	SiO_2			
Cation	Backwash	6.5	426	5	—	39	39	—
	HCl feed	2.5	8,550	1	—	12	—	12
	Displacement	1	92,177	0	—	21	—	21
	Rinsing (first 10 min)	0	4,790	8	—	13	—	13
	Rinsing	2	1,108	5	—	26	26	—
Anion	Backwash	3	771	1	—	8	8	—
	NaOH feed	6	25,270	2	—	17	—	17
	Displacement	12	166,367	3	—	26	—	26
	Rinsing (first 10 min)	13	9,610	6	—	11.2	—	11.2
	Rinsing	11	146	0	—	44.7	44.7	—
Polisher	Backwash	6	238	3.5	—	7	7	—
	NaOH feed	12	109,882	0	—	20	—	20
	Displacement	13	102,600	2	—	13	—	13
	Rinsing 1 (first 10 min)	12	1,135	2	—	11	—	11
	Rinsing	11	167	5.5	—	33	33	—
Total						301.9	157.7	144.2
Average Conductivity							476	—

Note: The conductivity of the used RO permit in this regeneration is 504 μs/cm².

Table 7-9 Ion Exchanger No. 2 Regeneration Analysis (Brief)

Unit	Sample	Lab Tests (sampled 5/22/2005)				Total Quantity (m³)	Reuse for Cooling Towers (m³)	Disposal (m³)
		pH	Conductivity	TSS	SiO₂			
Cation	Backwash	6	252	2	—	51	51	—
	HCl feed	7	5,494	0	—	13	—	13
	Displacement nos. 1 and 2	1	82,600	1	—	28	—	28
	Rinsing (first 10 min)	2	5,200	0	—	13	—	13
	Rinsing	3	861	0	—	33	33	—
Anion	Backwash	3	529	1	—	10	10	—
	NaOH feed	2	3,903	4	—	7	—	7
	Displacement nos. 1 and 2	10	82,553	0	—	30	—	30
	Rinsing (first 10 min)	12	1,291	0	—	11	—	11
	Rinsing	11	85	1	—	34	34	—
Polisher	Backwash	3	323	2	—	10	10	—
	NaOH feed	12	72,233	2	—	15	—	15
	Displacement	12	99,200	0	—	15	—	15
	Rinsing 1 (first 10 min)	12	1,771	0	—	13	—	13
	Rinsing	11	165	2	—	78	78	—
Total						361	216	145
Average Conductivity							369	

Note: The conductivity of the used RO permit in this regeneration is 285 µs/cm².

The First 10 Minutes of the Rinsing Stage

The conductivity of this water is not high if we compare it with the generated water from regenerant injection or displacement stages, and since it contains some of the regenerant (NaOH, HCL), it could be reused as feedwater for the regenerant injection step instead of the degasified water. This would also save some regenerant. However, since there is no space in the utilities section and the cost of tanks and fittings are high, this recommendation cannot be implemented.

The New RO Unit and Its Effect on the Ion Exchangers

Since the conductivity of the new RO is lower than the old unit, this will lead to an increase in time between the ion exchange regenerations and decrease the generated wastewater from ion exchangers regeneration. This will result in a cost savings in water and chemicals usage. Over time (within two years) the effectiveness of the new RO will decrease. This will lead to an increase in the RO conductivity. Hence, the refinery should do this recommendation to conserve water in the long run.

Savings from Reusing the Regeneration Effluent

In 2004, the cycle times for both units was 135 and the used water for regeneration process is 44,125.2 m^3. If we assumed that 44,125.2 m^3 is the annual water consumption for this purpose, that means the refinery could reuse around 24,612 m^3/year. Table 7–10 provides the 2004 ion exchangers regeneration cycle times and the assumed savings.

Also we will assume that the cost of this water is similar to the RO permit, because of its low salinity. The cost of the RO permit per cubic meter is $1.13.

The total savings is around $27,732/year, but the refinery needs to install a pump for the southern pool to reuse this water as cooling tower makeup and fitting to send the sand filter backwash water with the RO brine. This would cost around $704. Therefore, the payback period is less than half a month.

Table 7–10 2004 Regeneration and the Assumed Savings

Items	Unit No. 1	Unit No. 2	Total	Water Savings (m^3)	Money Savings (JD)
Number of regenerations	78	57	135	—	—
Total quantity of regeneration water (m^3)	23,548.2	20,577	44,125.2	—	—
Reused water (low TDS; m^3)	12,300.6	12,312	24,612.6	24,612.6	19,690.08
Disposal water (m^3)	11,247.6	8,265	19,512.6	—	—
Total savings				24,612	19,690

Note: Money savings are in Jordanian dollars.

Ion Exchangers Maintenance Program

The silica meter is out of use (see Figure 7–28), which indicates that there is no silica leak monitoring from the ion exchangers. Silica leakage leads to an increase in boiler blowdown rate and the corresponding chemicals and energy loss; therefore the silica should be monitored. The refinery needs to have a preventive maintenance program to ensure that instrumentations remain functional and reliable.

CP/P2 Opportunities for the Residential Area Water Consumption

There is a residential area near the refinery for employees and their families. The total number of residents is approximately 479, and the area is supplied with well water. According a carefully performed material balance, this residential area consumes 12.7m^3/h from the well for personal service, in addition to other points supplying RO water for drinking purposes. There are no individual flow meters.

This consumption means that the share per capita from the well water is 636 L/d/capita (0.636 m^3/d/capita). In Jordan, the average water consumption per person does not exceed 120 L/d/capita. This is inefficient use of at least 500 L/d/capita (0.5 m^3).

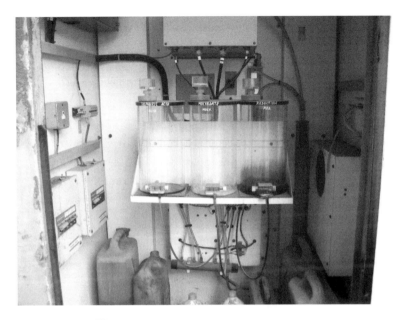

Figure 7–28 Silica meter.

CP/P2 Actions

It is recommended to provide the residential area with flow meters and change the current management practice for water supply and provide 30 m³ of water for each house without fee. Any excess use of water should require payment to encourage the resident to conserve and better use the water.

The total amount of water that could be saved would reach 204 m³/day thus 74,460 m³/year of well water with savings of $26,218/year ($0.35/m³).

CP/P2 Case Study at the Chevron Products Company, Richmond Refinery, United States[6]

The major waste streams generated at the Chevron Richmond Refinery include:

[6]*California Petroleum Refinery Hazardous Waste Source Reduction 1998 Assessment Report* (Sacramento: California Environmental Protection Agency, January 2004).

- API separator sludge generated during the separation of solids from refinery process waters at the API separators. Tank water solids, rust, scale, and other process equipment contaminants as well as soils from runoff make up this waste stream.
- Primary and secondary sludge generated when solids from various process equipment, such as tanks and process units, enter the process wastewater system and are removed prior to entering the API separators. Secondary sludge is generated after flocculants are added to an oil/water emulsion following primary treatment to aid in further separation of oil, water, and solids from the waste stream.
- Spent refining catalyst is generated from processes that treat, crack, or reform hydrocarbon streams by passing these streams over a metal-impregnated catalyst. The metals, present in the catalysts to enable them to achieve the necessary reactions, cause the catalysts to be hazardous when spent.
- Spent activated carbon is recovered downstream of the desalter during the treatment of desalter effluent. In the desalter, crude feed is washed with water to remove salts and sediments. The crude is then separated from the wash water (effluent). This effluent must then be treated to remove dissolved hydrocarbons (primarily benzene). Chevron Richmond uses granular activated carbon units in the final stage of effluent treatment to remove the benzene in the wash effluent. Captured benzene and other hydrocarbons may render the spent carbon hazardous.

The following are P2 projects that were implemented by Chevron aimed at source reduction of the hazardous waste streams.

CP/P2 Actions for the API Separator Sludge and Primary/Secondary Sludge

Chevron Richmond implemented two source reduction measures: (1) send oily sediments to Chevron's sister refinery in El Segundo, California, as feed for its coker, and (2) separate calcium-bearing wastewaters from phosphate-bearing wastewaters to prevent formation and deposition of calcium phosphate solids in the effluent treatment system.

Prior to 2000, Chevron Richmond processed oily sediments by filtering out solids to recover oil. The refinery shipped filtered solids off-site for hazardous waste incineration. Chevron developed a plan to transfer oily sediments to the company's refinery in El Segundo, California, as a coker feedstock. By recycling these oily sediments, Chevron eliminated oily sludge as a hazardous waste stream. Chevron estimated that coker sludge processing would reduce the oily sludge by approximately 80%.

The company's second measure involved separation of calcium-bearing wastewater from phosphate-bearing wastewater to prevent formation and deposition of calcium phosphate solids in the effluent treatment system. This resulted in an annual reduction of 20%.

CP/P2 Actions for the Spent Refining Catalyst

The company reported two activities: (1) regenerate and reuse spent hydrotreating catalyst, and (2) eliminate the cement processing of self-heating catalyst.

Spent diesel hydrotreating catalyst, spent naphtha hydrotreating catalyst, and spent jet hydrotreating catalyst can be sent off-site for regeneration. The catalysts can then be reused on-site or by another facility. While the same amount of catalysts are generated, they are shipped off-site for regeneration instead of for disposal. The refinery estimated 8% of the total catalyst waste stream can be regenerated and reused. This measure is off-site recycling and reduces waste disposal.

Certain catalysts are self-heating. Chevron's prior practice of processing self-heating catalyst with cement to eliminate the risk of self-heating during accumulation and transportation resulted in high disposal costs. Chevron Richmond started sealing the self-heating catalyst in transportation containers with dry ice to provide a carbon dioxide (CO_2) blanket and oxygen-deficient atmosphere. This measure reduced the risk associated with handling and transporting self-heating catalysts and without the addition of cement and, thus, decreased the amount of waste sent off-site for disposal. The facility closely supervised the initial shipments sent without cement processing to assess any self-heating and safety concerns. Chevron determined that the use of CO_2 and improved sealing of transportation containers effectively minimized the safety concern of self-heating. This activity resulted in an annual reduction for self-heating catalyst of approximately 30%.

CP/P2 Case Study at the Beijing Yanshan Chemical Corporation, China[7]

The Beijing Yanshan Chemical Corporation (BYPC), with 5,800 employees, was founded in 1969. A major part of the enterprise is its refinery, which consists of 16 processing units including crude distillation and fluid catalytic cracker (FCC), which have a total throughput capacity of 6.6 million tons of crude oil annually. The product line of the refinery includes gasoline, diesel, kerosene, lubes, and wax. The refinery participated in the Sino-Norwegian cleaner production program in China. The following is a summary of the CP/P2 opportunities that the refinery captured.

CP/P2 Opportunities for the FCC

The FCC unit 2 (FCCU2) is a major process unit that was placed into service in 1983. In 1985, it was upgraded to process vacuum residual (at a capacity of 800,000 tons of feed a year). The feeds to the unit are heavy AGO and VGO, which are cracked from catalyst, out of which lighter oil goes through the distillation tower and stabilizer to produce the gasoline and diesel. The spent catalyst is regenerated and reused after coking. The major emissions from the unit are flue gas from regenerator and furnaces, waste alkali from gasoline and LPG washing, and wastewater containing oil and sulfur.

---

[7]UNEP IE Case Study by the Sino-Norwegian Cleaner Production Program (1997), Office of Sino-Norwegian Cleaner Production Program—China International Training Center for Sustainable Development, 109 Wanquanhe Road, Haidian District, Beijing 100080, P. R. China (Telephone: 86-10-2588146).

CP/P2 Actions

After screening CP/P2 opportunities identified from an audit, a total of 28 recommended actions were identified and evaluated. Ten of the options fell into the category of no cost/low cost and were immediately implemented. Fourteen of the options fell into the medium-cost category, all showing reasonable ROIs from a cost-benefit analysis. Among the medium-cost investments, three were reported to have been implemented immediately. An additional 11 investments fell into the high-cost category and required further assessment.

No-cost/low-cost opportunities jumped out from observing poor housekeeping practices. The recommended actions included

1. Monitor water usage.
2. Keep contact with the environment monitor section to control the waste discharge.
3. Maintain operation inspections to prevent leakage of materials.
4. Motivate employees to take part in the CP/P2 program.
5. Extend the process operation period to decrease the shutdown time.
6. Optimize the operation to reduce pollutant content in the emission gas from furnaces.
7. Stress equipment maintenance to prevent damage that leads to leakage.
8. Recover residual materials in the equipment to be repaired.
9. Recover residual materials in the equipment and pipeline at shutdown time.
10. Recover the used lubes for pumps.

Among the medium-cost recommendations were

1. Perfect the catalyst recovery system of the third cyclone.
2. Install a dust removing device on the pipe of flue gas regenerator.
3. Modify the ammonia recycle pipe in the ammonia injecting system.

4. Install a one-way valve on ammonia injection distribution.
5. Change the oil seal into a mechanic seal for the air blower to eliminate lube pollution.
6. Install a bypass pipe to the metering gauge of the ammonia injection.
7. Send demercaptan end gas to the flare system.
8. Analyze product quality with an online instrument and connect it to the computer system.
9. Install an alkaline drainage pipe to separate the alkaline drainage and oily water flow.
10. Remove the process of LPG wash to eliminate LPG waste alkali.
11. Recover heat in the flue gas regenerator with a boiler.
12. Improve the air cooling system to recover soft water.
13. Neutralize waste alkali with regenerator flue gas.

Among the high-cost alternatives were

1. Develop a high-efficiency carburant to lower the CO in the flue gas regenerator.
2. Develop a new catalyst to reduce the catalyst consumption.
3. Develop a new technology for catalyst regeneration.
4. Develop a new technology for steam atomizing to reduce sulfide waste water.

The enterprise improved both environmental and financial performance from those opportunities it invested in. Some of the financial savings that are reported here are in U.S. dollars. A currency equivalent of $0.120 per yuan RMB (people's currency) was used.

Improved performance was achieved in the following areas:

- *Energy.* While there were no benefits from the no-cost/low-cost actions implemented, the medium-cost investments reduced steam demand from 22 ton/h to 11 ton/h or about a 50% savings.

- *Water demand.* The low-cost/no-cost actions implemented resulted in a modest savings by reducing water use from 17.0 ton/h to 16.2 ton/h, or a 4.7% reduction. By implementing several of the medium-cost recommendations water demand was reduced to 11.0 ton/h for an impressive 35.3% savings.
- *Wastewater and solid wastes.* The no-cost/low-cost actions implemented resulted in a 13,200 ton/yr reduction in wastewater generation (a 9% reduction). Accompanying this was a reduction in chemical oxygen demand (COD) by 3.1 ton/yr for an impressive 32% reduction. Solid waste was reduced by 50 ton/yr or 10%. Pollution fees were reportedly reduced by $94,800. This represents a savings that continues year after year and frees up capital for other improvements that make the enterprise more competitive.

The facility reported that, over the last several years, it invested $1.22 million in CP/P2 projects identified from the audit and captured annual net savings of $825,600. The annual steam demand alone has been reduced by 77,000 tons. The internal rates of return for many individual projects implemented ranged from >50% to as high as 80% with attractive payback periods of under two years.

CP/P2 Case Study at the Indian Oil Corporation Limited, Gujarat Refinery, Vaodara, India[8]

This refinery was commissioned in 1963 and presently has a capacity of 13.7 MMTPA (million metric tones per annum). A block diagram of the facility is shown in Figure 7–29. This facility is registered under the ISO-14001 certification for environmental management.

The refinery has focused on energy conservation opportunities. Energy consumption is monitored on a daily basis and demand optimization is an ongoing activity. The efforts of a dedicated team aided the Gujarat Refinery to bring down the energy consumption

---

[8]Available at various WWW page reports; see http://www.bee-india.nic.in/sidelinks/EC%20Award/Current.html.

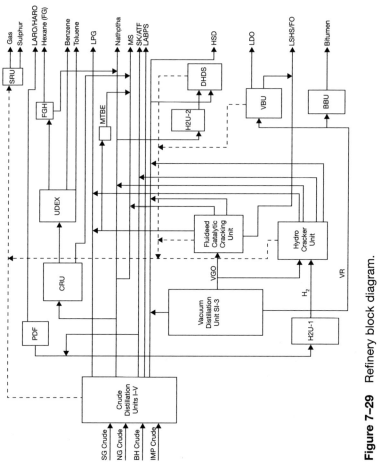

Figure 7-29 Refinery block diagram.

level from 110.4 MBTU/BBL/NRGF in 2000 to 101.4 MBTU/BBL/NRGF in 2004. The following are summaries of major achievements and innovative approaches applied.

CP/P2 Achievements

The following are but a few of the projects that have been or are being implemented and focus on P2 and resource conservation:

- The refinery has become a zero-discharge facility. Its entire treated effluent is recycled back for firewater and cooling water makeup.
- Total SO_2 emissions have been maintained below the stipulated limit. This is an example of a facility recognizing that there are benefits to going beyond the minimum legal requirements.
- The Energy and Resources Institute (TERI), New Delhi, and the Indian Oil Corporation Limited (IOCL) R&D Centre, Faridabad, developed a new bioremediation technique for degradation of oily sludge. They developed a bacterial system known as *Oilivorous-S* to biodegrade oily sludge and sulfur containing hydrocarbons, leaving behind no harmful effects. The facility successfully treated 1,750 megatons (MT) of oily sludge using the bioremediation technique. Treatment of a second lot of 1,700 MT oily sludge is in progress.
- The major fuel consumers in a refinery are process heaters. To reduce radiation losses through heater walls, proper installation and maintenance of refractory is crucial. The emissivity of a refractory decreases with a rise in temperature, which in turn increases energy consumption. High emissive ceramic coatings increase the emissivity of heater wall refractories, resulting in a reduction of fuel consumption. The major benefits of applying high emissivity coatings are
  - It increases the emissive power of the refractory of heater walls and reflects back the heat, resulting in fuel savings.
  - It increases the life of refractory, preventing the formation of hot spots.
  - It can withstand temperatures up to 1,500°C.

- It enhances productivity with consistent quality due to a stable skin temperature.

This technology was successfully applied at the refinery, resulting in significant fuel savings.

- A project known as Preheat Improvement in Feed Preparation unit-1 (FPU-1) was implemented. The preheat depends on the quantity of hot feed. Processing more hot feed increases the preheat, resulting in lower fuel consumption in the process heater. Earlier feed of FPU-1 consisted of mixed RCO (cold and hot) ex atmospheric unit no 4. To increase the hot feed quantity, facilities for processing hot feed (RCO) from AU-5 were commissioned. An increase in preheat temperature of 8°C resulted in fuel savings of 950 MT per year.

- A project was implemented to replace normal refrigeration for the control room cooling with vapor absorption. A vapor absorption system is environmentally friendly and energy efficient, since no CFCs are emitted, as in the case of conventional compression chillers. These systems operate on low-pressure steam, which is readily available in any refinery. This concept saves on costly electricity. Vapor absorption technology was installed in the plant's laboratory at a cost of 30 lakhs. Approximately 943,000 KWH of power per year of savings has been projected from the application of this technology.

- The refinery's FCC unit was bottlenecked at the high-temperature end as the CO boiler was operating at its MCR of 78 tons/hr. By reducing the burner tips of the CO boiler, the fuel oil required for auxiliary firing was reduced. The redesign reduced fuel oil demand by the boiler to 13 MTPD from 25 MTPD and enabled the FCC to operate at higher throughputs.

- The air supply to the regenerator in the refinery's FCC unit is met by a turbo-driven main air blower. The consumption of medium-pressure steam to the turbine is a function of the air requirement. The addition of a CO promoter reduced the afterburn and, hence, reduced the air requirement of the regenerator. On addition of the additive, the requirement of air to the regenerator was reduced by 1,500 Nm^3/hr at the same throughput. This resulted in a savings of 2 MTPH of medium-pressure steam consumption.

- A crude pump based on an old design using gland packing was prone to leakage and fugitive emissions. The pump also represented a safety hazard. The pump's gland packing was replaced with a mechanical seal, thereby stopping crude oil losses. It also reduced frictional losses compared to the gland packing and saved power.

- Floating roofs were designed more than 100 years ago as a safety precaution to keep explosive gasses from building up between the roof of a tank and its petroleum contents. Today, the emphasis is on controlling losses, atmospheric pollution, as well as safety. To arrest vapor losses from floating roof tanks, floating roof tanks were provided double seals, that is, a secondary seal in addition to existing seals.

- A study was commissioned to determine the condition of steam traps and steam leaks throughout the refinery. The survey identified damaged or leaky steam traps and leaky steam points and placed an aggressive preventive maintenance program into effect. The total repair or replacement of steam traps and attending to the steam leaks as per recommendation of the task force was given high priority. This resulted in saving of Rs. 162 lakhs in fuel cost.

Energy conservation in any refinery plays a vital role toward cost reduction as well as good environmental performance. CP/P2 opportunities in this area are directly related to fuel savings while reducing the adverse effects of fuel burning on the environment.

CP/P2 Case Study at California Refineries

The California Petroleum Refinery Hazardous Waste Source Reduction 1998 Assessment Report (Sacramento: California Environmental Protection Agency, January 2004) is an excellent resource for examples of CP/P2 projects and programs implemented by different refineries. Tables 7–11 and 7–12 highlight some of the improved environmental performances achieved by select refineries.

Table 7–11 California Petroleum Industry Aqueous (Category A) Waste Reduction Results (1994 versus 1998 Waste Generation)

Site	1994 Quantity (lbs.)	1998 Quantity (lbs.)	Amount Changed (lbs.)	Amount Changed (%)
Martinez Refining Co., a Division of Equilon Enterprises	14,400,000,000	7,890,000,000	6,510,000,000	45.2
Tosco Refining Co., San Francisco Refinery Carbon Plant	8,716,050	0	8,716,050	100.0
Golden Eagle Refinery	2,203,116,000	1,759,009,700	444,106,300	20.2
Chevron U.S.A. Inc., El Segundo Refinery	129,990	26,642	100,348	77.2
Tosco Refining Co., Santa Maria Refinery and Carbon Plant	376,700	0	376,700	100.0
Totals	16,612,338,740	9,649,036,342	6,963,299,398	41.9

Table 7-12 California Petroleum Industry Nonaqueous (Category B) Waste Reduction Data (1994 versus 1998 Waste Generation)

Site	1994 Quantity (lbs.)	1998 Quantity (lbs.)	Amount Changed (lbs.)	Amount Changed (%)
Chevron Products Co., Richmond Refinery	13,934,000	10,912,000	3,022,000	21.7
Valero Refining Co.	83,347,800	8,382,800	74,965,000	89.9
Chevron U.S.A. Inc., El Segundo Refinery	67,880,000	49,255,336	18,624,664	27.4
Tosco Refining Co., Carson, Wilmington, and Marine Terminal	55,905,703	5,872,955	50,032,748	89.5
Paramount Petroleum Corp.	1,521,680	146,240	1,375,440	90.4
Valero Refining Co., Wilmington Asphalt Plant	152,000	96,100	55,900	36.8
Oildale Refinery	1,434,948	668,390	766,558	53.4
Tosco Refining Co., Santa Maria Refinery and Carbon Plant	2,719,490	1,270,050	1,449,440	53.3
Totals	226,895,621	76,603,871	150,291,750	66.2

Index

Adjusted internal rate of return, 16, 21, 24
Air pollution controls, 37
American National Standards Institute, 52
Ammonium Perchlorate, 31
Annual recurring costs, 18
API separator opportunities, 285
Attitudinal changes, 81
Auto painting case study, 151

Backwashing, 280–281
Beijing Yanshan Chemical Corp., case studies, 296–299
Benchmarking, 55
Best available technologies, 8
Biochemical oxygen demand, 172
Biosolids, 22
Boiler blowdown, 264–270
British Standard 7750, 72–74

California refineries, case studies, 303–305

Capital equipment costs, 5
Cash flow diagrams, 19
CDM International, 207
Checking and corrective actions, 62
Chemical oxygen demand, 172
Chevron Products Co., 293–296
Cleaner production, 79, 111
Community relations, 54
Compliance requirements, 26
Comprehensive Environmental Response, Compensation and Liability Act (CERCLA), 41
Condensate recovery systems, 270–275
Conformity assessment, 51
Continual improvement cycle process, 49
Continual improvement cycle, 47
Cooling towers, 282–284
Cost categories, 16
Cost-benefit analyses, 8
Crude distillation, 212
Crude unloading stations, 214

CSX Corp., 43
Current dollar value, 15

Destructive Removal Efficiency, 35, 40
Direct costs, 4
Direct labor costs, 5
Discount factors for LCCA, 18
Discounted payback calculation, 25
Discounted payback, 22
Discounting, 14

Eco-efficiency, 78, 143
Eco-management and audit scheme (EMAS), 66–72
Economic performance factors, 21
EMAS implementation route, 72
EMIS design options, 128
EMIS implementation strategies, 127
EMIS informational flows, 124
EMIS organizational activities, 123
EMS certification infrastructure, 53
EMS elements, 48
EMS implementation, 60
Energy usage, 167
Engineering economic analysis, 3
Environmental aspects, 136–137
Environmental auditing, 50
Environmental cost accounting, 3
Environmental declarations, 50
Environmental forensics, 2
Environmental goals, 57
Environmental impact statements, 5
Environmental labels, 50
Environmental management hierarchy, 151
Environmental management information systems (EMIS), 121–128
Environmental management initiatives, 119–121

Environmental management system definition, 45
Environmental management system (EMS), 1
examples of, 65
Environmental performance evaluation, 50
Environmental performance targets, 57
Environmental policy checklist, 58
Environmental policy statement, 56
Environmental policy, 47
Environmental Protection Agency, 74–75
Environmental stewardship, 42
European Union Council, 66
Exxon/Mobil Corp., 66

Financial calculation tools, 3
Fluid catalytic cracking, 212
Ford Motor Co., 156
Fuel additive programs, 39

Gap analyses, 125
Gaseous emissions, 173
Gasoline leakage, 219
General Motors Corp., 43, 142
Grease and oil, 172
Green productivity, 78

Health and safety assessments, 5
Health and safety effects analyses, 56
Hewlett-Packard, 142
Hidden costs, 4, 5
Hormel Food Corp. 43
Hydrotreating, 212

IER audit questionnaire, 90–110
IER case study, 207

Index

IER toolkit, 87
Illinois Central Railroad, 42–45, 83
Incinerators, 247
Indian Oil Corp. case studies, 299–303
Indirect costs, 5
Industrial ecology, 78
Inflation, 14–15
Initial environmental review (IER), 85–86
In-plant assessments, 148
Integrated EMS/P2 model, 116
Integration of EMS and P2, 113–121
Interest, 14
International Chamber of Commerce's Business Center for Sustainable Development, 66
Ion exchangers, 284
ISO 14000 standard, 65
ISO 14001 clauses, 51
ISO 9000 quality standards, 51

Johnson Controls, 142
Jordan Petroleum Refinery, case study, 262

Koppers Co., 33–41

Landfilling, 249
LCC calculation methodology, 20–21
Leak detection and repair programs (LDAR), 234
Legal costs, 5
Life-cycle cost analyses, 8, 12–20
Life-cycle costing calculations, 11
Life-cycle costing, 4
Life-cycle perspective, 143
Life-cost planning, 11
Loading stations, 218

Lockheed-Martin Corp., 29–33
Long-term waste reduction options, 184
LPG recovery, 212

Management review process, 47
Management reviews, 63
Material balances, 163–181
Material Safety Data Sheets, 32
Materials management, 143
Mississippi Department of Environmental Quality, 35

NACE codes, 69
NAFTA, 66
Net savings calculation, 23
Net-savings measure, 21
Nongovernmental organizations, 66
Normal costs, 4
North American Commission for Environmental Cooperation, 66

Off-site waste disposal, 175–176
Operating cost reductions, 132
Operating, maintenance, and repair costs, 17
Operational Integrity Management System for EHS, 66

P2 pre-assessment steps, 158
P2 toolkit, 190–206
Petroleum product losses, 220
Planning, 58
Pollution control technology, 22
Pollution cost tiers, 9
Pollution fees, 4
Pollution prevention auditing steps, 158–189

Pollution prevention hierarchy, 83
Pollution prevention terminology, 74
Poor housekeeping practices, 222
Present value formulas, 18
Present value, 14

Quality control, 42

Raw material losses, 159
RCRA definitions, 38
RCRA regulated hazardous wastes, 27
RCRA, 30, 76
Recycling, 75
Redlands case study, 29
Registrar Accreditation Board, 52
Remedial action costs, 6
Resource efficiency, 144
Responsible Care Program, 52
Revenues, 19
Reverse osmosis, 279
Root causes, 137–140

Savings-to-investment ration, 16, 21
Segregation, 183
Simple payback, 22
Single present value factor, 17
Slop tanks, 256–257
Sludge pond management, 240–244
Smartwood EMS, 65
Sour water stripping, 245
Source reduction, 75–76
Stakeholder confidence, 63
Steam losses, 236, 275–279
Steam recovery, 262

Supply chain environmental risk management, 144–145
Supply chain greening, 135
Suspended solids, 172

Tank bottom sludge, 235
Tetraethyl lead, 246
Third-party damages, 6
Threshold Limit Values (TLVs), 100
Time value of money, 11
Time weighted average (TWA), 100
Total quality environmental management, 154–155
Toxic use reduction, 77
Trichloroethylene (TCE), 30–31, 155–158

Underground storage tanks, 10
Unit operations, 160
United States Agency for International Development, 207

Vacuum distillation, 212

Waste minimization programs, 45
Waste minimization, 76
Waste processing, 44
Waste reduction, 77
Waste reuse and recycling, 169–170
Water usage, 168–169
World Wide Web, 2

Zarqa industrialized zone, 212